# Brain-Computer Interface and Intelligent Transportation

# Brain-Computer Interface and Intelligent Transportation

**Guanghui Zhao**
Guizhou University of Finance and Economics, China

World Scientific

NEW JERSEY • LONDON • SINGAPORE • BEIJING • SHANGHAI • HONG KONG • TAIPEI • CHENNAI • TOKYO

*Published by*

World Scientific Publishing Co. Pte. Ltd.
5 Toh Tuck Link, Singapore 596224
*USA office:* 27 Warren Street, Suite 401-402, Hackensack, NJ 07601
*UK office:* 57 Shelton Street, Covent Garden, London WC2H 9HE

**Library of Congress Cataloging-in-Publication Data**
Names: Zhao, Guanghui, 1976– author
Title: Brain-computer interface and intelligent transportation / Guanghui Zhao,
Guizhou University of Finance and Economics, China.
Description: New Jersey : World Scientific, [2027] | Includes bibliographical references and index.
Identifiers: LCCN 2026001716 | ISBN 9789819829293 hardcover |
ISBN 9789819829309 ebook for institutions | ISBN 9789819829316 ebook for individuals
Subjects: LCSH: Intelligent transportation systems--Technological innovations |
Intelligent transportation systems--Technological innovations--China |
Brain-computer interfaces--Industrial applications | Artificial intelligence--Industrial applications
Classification: LCC TE228.3 .Z432 2027
LC record available at https://lccn.loc.gov/2026001716

**British Library Cataloguing-in-Publication Data**
A catalogue record for this book is available from the British Library.



For any available supplementary material, please visit
https://www.worldscientific.com/worldscibooks/10.1142/14774#t=suppl

Desk Editors: Eshak Nabi Akbar Ali/Veronica Lee

Typeset by Stallion Press
Email: enquiries@stallionpress.com

# Preface

Brain–Computer Interface (BCI), as an interdisciplinary technology, is gradually transforming various aspects of intelligent transportation. It combines human consciousness with computer technology, providing new possibilities for human–vehicle interaction, autonomous driving, and intelligent transportation. This book explores the relationship between brain–computer interfaces (BCIs) and intelligent transportation, analyzes their impact on intelligent transportation, and introduces the applications and operations of BCIs in the transportation field.

The rapid development of BCI technology has profoundly changed people's cognition and usage of transportation tools. In the past, drivers needed to operate the steering wheel with both hands and control the accelerator and brake with their feet. Now, BCIs enable drivers to control vehicle movement through their thoughts, making driving more convenient and safer.

The introduction of BCI technology has greatly enhanced the intelligence level of transportation tools. Through BCIs, vehicles can perceive the driver's intentions in real time, automatically adjust driving status, and avoid traffic accidents caused by human operational errors. At the same time, BCI technology also improves the operational efficiency of transportation tools and reduces congestion.

Currently, the applications of BCIs in the transportation field mainly include human–vehicle interaction and autonomous driving. In terms of human–vehicle interaction, drivers can convey driving intentions to the vehicle through BCIs, while the vehicle can perceive the driver's emotions and physiological states via BCIs, achieving more humanized services. Regarding autonomous driving, BCI technology can enable vchiclcs to havc highcr lcvcls of autonomous driving capabilitics, such as fully automated driving on highways or urban roads.

The application of BCI technology in the field of intelligent transportation holds enormous potential. It can not only improve transportation safety and efficiency but also enhance people's travel experience. In the future, with the continuous development and improvement of BCI technology, we have every reason to believe that it will play an even more important role in intelligent transportation.

# About the Author

**Guanghui Zhao** (born September 1976) is a Professor at Guizhou University of Finance and Economics, China. He is also a Researcher and Senior Visiting Fellow at the School of Engineering, University of Michigan, USA. In addition, he is the Founding Director of the Research Center for Modern Transport Development at the College of Management Cadres, Ministry of Transport, as well as a Researcher at the Research Center for Industrial and Academic Research Issues, Wuhan University.

Prof. Zhao has led numerous major projects, including National Social Science Fund of China (NSSFC) research projects commissioned by the Ministry of Transport, the China Postdoctoral Foundation Special Fund Project, Special Fund Projects from the China Postdoctoral Foundation, and over 30 China National Academy of Educational Sciences planning projects. He has authored over 100 academic papers in leading core journals and more than 20 treatises.

Prof. Zhao has published 10 monographs on transportation science and technology, including *Big Data Transportation Systems* (World Scientific), *Assessment and Policy Research on the Transportation Science and Technology Development in China* (World Scientific), *Redefining Traffic: How AI Leads the Change* (World Scientific), and 區塊鏈交通：開創未來交通新格局 (English title: *Blockchain in Transportation: Creating a New Landscape for Future Transportation*) (Posts & Telecommunications Press).

# Contents

# Part I

# Basic Layer of the Upper Volume

# Chapter 1

# Overview of Brain–Computer Interface Technology

## 1.1 Definition and Development History of Brain–Computer Interface

Brain–Computer Interface (BCI) is a technology that directly establishes a communication connection between the brain and external devices. It allows users to control devices or interact with computer systems solely through brain activity, without the need for traditional muscle or voice input.

The emergence of BCI technology undoubtedly opens a brand-new door for the future of humanity. Its application fields are broad and profound, ranging from healthcare to daily life, and even cutting-edge scientific research.

In the healthcare field, BCI technology brings new hope to patients who have lost mobility for various reasons. Imagine patients who are unable to move due to stroke, spinal cord injury, or muscular diseases; they can now regain mobility and return to normal life by controlling prosthetics, wheelchairs, or other assistive devices solely through brain activity via BCI technology.

In daily life, BCI technology also brings unprecedented convenience. For example, through BCI technology, people can control smart home devices such as lights, air conditioners, and televisions solely through brain activity, achieving true "mind over matter." Additionally, BCI technology can be applied in education, entertainment, and other fields, providing people with richer and more diverse experiences.

In cutting-edge scientific research, BCI technology holds enormous potential. By real-time monitoring and analyzing brain activity, scientists can gain deeper insights into the working principles of the human brain, thereby advancing related fields such as neuroscience and cognitive science. Moreover, BCI technology can be used to study complex psychological phenomena such as human consciousness, memory, and emotions, offering more clues to unravel the mysteries of the brain.

Of course, the development of BCI technology also faces many challenges and difficulties. For example, how to ensure data accuracy and security? How to reduce device costs and complexity? How to guarantee user privacy and ethical standards? These issues require continuous exploration and resolution.

If we review the development history of BCI, we can roughly divide its timeline into five stages based on research and development periods, as shown in Figure 1.1.

**1. Early Exploration (1960s–1970s)**

During this period, scientists began exploring direct connections between brain electrical activity and external devices. Early BCI research mainly focused on animal experiments, controlling simple external devices by recording brain electrical signals.

Over time, research in the BCI field gradually shifted from animal experiments to more complex human applications. Scientists continuously explored the mysterious connection between the brain and machines, seeking new ways to translate human consciousness into practical actions.

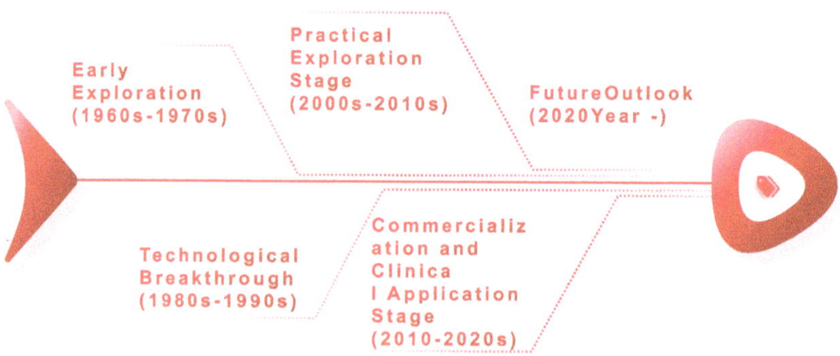


**Figure 1.1.** Development history of BCIs.

The initial challenge in human BCI research was how to accurately read and interpret the brain's electrical signals. The brain is an extremely complex neural network, and the electrical signals it produces are often elusive. However, with technological advances, researchers developed more sophisticated electrodes and algorithms capable of capturing and interpreting finer brain electrical activities.

In 1964, Joseph Kamiya at the University of California, Los Angeles, first demonstrated that humans could consciously control brainwaves, pioneering research in brainwave training.

In 1969, Eberhard Fetz proved in monkey experiments that animals could learn to control neural activity, an important precursor to the BCI concept.

During this process, scientists discovered that different brain regions correspond to different functions and tasks. For example, the motor cortex is closely related to body movement control, while the visual cortex is responsible for processing visual information. By precisely locating these brain regions and interpreting the electrical signals they generate, researchers could better understand the brain's working mechanisms and develop more accurate BCI systems.

As research deepened, the application areas of BCI systems gradually expanded.

**2. Technological Breakthroughs (1980s–1990s)**

With the development of computer science and signal processing technology, it has become possible to extract patterns related to mental tasks from EEG signals. Research during this period focused on developing systems capable of interpreting and utilizing these signals.

Driven by this wave, researchers have not only focused on improving the accuracy of signal decoding but also strived to broaden its application scenarios. In 1988, Jonathan Wolpaw and others proposed controlling a computer cursor through EEG signals, marking a key advancement in early BCI systems.

In the 1990s, research found that the P300 EEG response (an event-related potential elicited by oddball tasks) could serve as an effective control signal for BCIs. During this stage, the research focus shifted to experimental validation, gradually demonstrating practical application possibilities by decoding EEG and single-neuron recording data, such as controlling a computer cursor.

Of course, EEG signal decoding technology still faces many challenges. First, due to the complexity of brain activity, accurately decoding EEG signals requires highly specialized knowledge and techniques. Second, the acquisition and processing of EEG signals may be affected by various factors, such as electrode placement and scalp oils, all of which impact the accuracy of decoding results. Finally, how to apply these technologies in real-world scenarios while ensuring their safety and reliability remains a problem researchers need to address.

**3. Practical Exploration Stage (2000s–2010s)**

During this stage, BCI technology began transitioning from the laboratory to practical applications. In 2004, Matthew Nagle became the first quadriplegic patient to operate a computer for basic tasks through a BCI system, as part of the BrainGate system. In 2006, researchers successfully used EEG data to control a cursor in a virtual environment, greatly expanding the potential applications of BCIs.

In this stage, BCI technology gradually moved closer to practical use, with more functional systems emerging, such as the BrainGate system. Both invasive and non-invasive BCI technologies developed rapidly.

With the continuous advancement of BCI technology in practical applications, its potential has been progressively explored and amplified. Today, BCI technology is no longer limited to assisting disabled individuals; it has begun to penetrate broader fields, bringing unprecedented convenience to human daily life and work.

**4. Commercialization and Clinical Application Stage (2010s–2020s)**

During this stage, the practical application scope of BCI technology continued to expand, and the technology further developed. In 2013, Emotiv launched a consumer-grade EEG headset, marking the entry of BCI technology into the consumer market.

In 2020, Neuralink, led by Elon Musk, released a BCI system based on high-throughput neural probes, featuring high-resolution and multichannel signal acquisition capabilities.

At this stage, BCI technology has ushered in opportunities for commercialization and scaling. The lightweight and portable hardware, along with the introduction of machine learning algorithms, have significantly enhanced the practicality and user experience of BCI systems.

In the medical field, BCI technology is being used to diagnose and treat various neurological disorders. For example, by monitoring and analyzing brain activity, doctors can more accurately diagnose neurological diseases such as epilepsy and Parkinson's disease, and accordingly develop more precise treatment plans. At the same time, BCI technology can assist in treating these diseases by simulating normal neural signals, providing patients with more effective therapeutic methods.

In the education sector, BCI technology has also demonstrated tremendous application potential. By real-time monitoring of students' learning states, teachers can better understand students' learning needs and difficulties, thereby offering more personalized teaching plans. Additionally, BCI technology can help students better manage their study time and attention, improving learning efficiency.

In the entertainment field, BCI technology likewise holds broad application prospects. By controlling game characters or virtual environments through brain activity, players can gain a more realistic and immersive gaming experience. Moreover, BCI technology can adjust game content and difficulty based on players' psychological states and emotional changes, making games more challenging and engaging.

It is foreseeable that with the continuous development and improvement of BCI technology, it will demonstrate its unique value and potential in more fields. In the future, we expect to see more innovative applications based on BCI technology, which will bring humans a more convenient, efficient, and intelligent lifestyle.

**5. Future Outlook (2020–)**

Currently, BCI technology is still evolving. Future development directions may include improving system portability and user-friendliness, as well as expanding its applications among healthy populations, such as enhancing efficiency in gaming, education, and workplaces.

In 2021, Neuralink demonstrated a wireless BCI device that allowed laboratory monkeys to control games using brain signals. In 2022, more research teams attempted to combine BCI technology with other technologies such as AI and edge computing to improve system real-time performance and accuracy.

At the current stage, BCI technology is advancing toward more cutting-edge directions, integrating wireless transmission technology, fully invasive systems, edge computing, and artificial intelligence, striving to achieve more efficient and less invasive BCI systems in the near future, and exploring the possibilities of future technologies such as brain–cloud interfaces.

Overall, the development history of BCI technology reflects the transition from basic research to practical applications, while also demonstrating great potential for improving human quality of life in the future.

## 1.2 The Working Principle of Brain–Computer Interface

The working principle of BCI is based on detecting, interpreting, and converting brain activity to achieve control over external devices or interaction with computer systems. The general working principle of a BCI system includes these components.

**1. Signal Detection and Acquisition**

First, the BCI system needs to detect the electrical signals generated by brain activity. This is usually achieved by placing electrodes on the scalp, which can record EEG signals. In some advanced BCI systems, implanted electrodes may be used to record signals more precisely.

Signal acquisition is a key step in BCI operation, involving the recording of neural activity from the brain. These signals can be obtained through the following methods:

(1) Invasive Methods

Intracortical recordings are a method that directly records neuronal activity by surgically implanting electrodes, as shown in Figure 1.2. This method can provide data with high temporal and spatial resolution, allowing in-depth analysis of neural system functions. However, it is important to note that due to the need for surgical implantation of electrodes, this technique carries a certain risk of infection.

(2) Non-invasive Methods

There are three common non-invasive methods, roughly illustrated in Figure 1.3:

(1) Electroencephalography (EEG) is a non-invasive technique that records electrical activity on the scalp by wearing an external electrode cap. This method is widely adopted due to its convenience and broad applicability.

(2) Functional Magnetic Resonance Imaging (fMRI) is an advanced imaging technology that indirectly measures brain activity by detecting changes in cerebral blood flow. Its notable feature is high resolution, but its real-time capability is relatively weak.
(3) Near-Infrared Spectroscopy (NIRS) uses changes in the absorption of near-infrared light in brain tissue to monitor brain activity, providing a new perspective for neuroscience research.

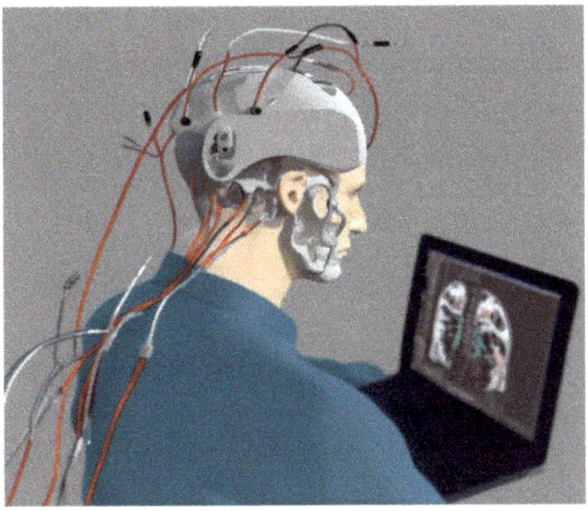

**Figure 1.2.** Illustration of invasive methods.

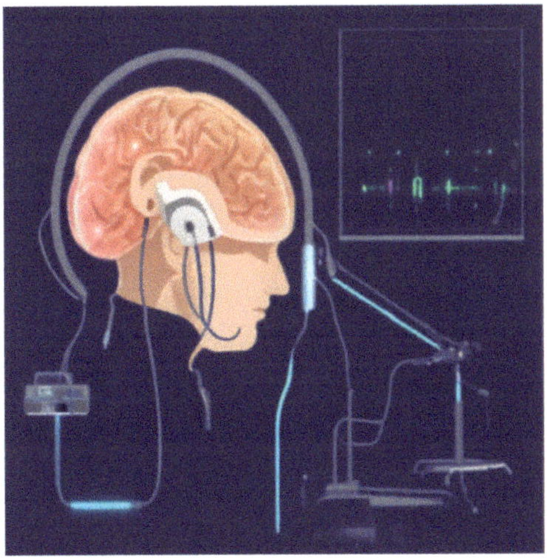

**Figure 1.3.** Non-invasive schematic.

**2. Signal Amplification and Filtering**

The detected EEG signals are very weak, so amplifiers are needed to enhance the signal strength. At the same time, to improve signal quality, filters are used to remove noise and other irrelevant signals, such as electrical signals generated by muscle activity.

**3. Signal Processing**

The amplified and filtered signals need to undergo complex signal processing to extract features related to the user's intentions. This includes steps such as feature extraction, pattern recognition, and classification algorithms. These algorithms can identify brain signal patterns associated with specific mental activities or commands:

(1) Feature Extraction

During signal processing and analysis, to accurately identify specific brain activities, it is necessary to precisely extract relevant information features from the preprocessed signals. Common feature extraction methods mainly include the following:

First, time-domain feature extraction directly utilizes the time series data of the signal to capture the dynamic characteristics of brain activity over time.

Second, frequency domain feature extraction allows us to extract frequency components from signals, such as alpha waves and beta waves, by applying methods such as Fast Fourier Transform (FFT), revealing specific patterns of brain activity in the frequency domain.

Finally, for spatial feature extraction, with the help of advanced methods such as Independent Component Analysis (ICA), we can further extract the spatial distribution characteristics of signals, thereby comprehensively analyzing the spatial layout and interaction patterns of brain activity.

(2) Pattern Recognition and Classification Algorithms

Pattern recognition technology aims to accurately decode the extracted features by applying machine learning algorithms and thereby identify the corresponding intentions or commands. Commonly used algorithms in the implementation process include the following:

(1) Linear Discriminant Analysis (LDA) is a linear method whose core function is to effectively separate different categories.
(2) Support Vector Machine (SVM) algorithm performs excellently in classification and regression problems and is widely adopted.

(3) Deep learning algorithms such as Convolutional Neural Networks (CNN) have unique advantages, especially when handling complex pattern recognition tasks.

**4. Command Translation**
The recognized EEG signal patterns are converted into control commands. These commands can be simple, such as moving a cursor or switching devices on and off, or more complex command sets used to control multi-degree-of-freedom robotic arms or wheelchairs.

**5. Device Control or Interaction**
The converted commands are sent to external devices such as computers, wheelchairs, prosthetics, or game controllers to achieve the user's goals. For example, users can control the movement direction of a wheelchair or type on a computer through their thoughts.

The working principle of BCI reflects the integration of multiple disciplines, including neuroscience, signal processing, computer science, and biomedical engineering. With technological advancements, the performance and applicability of BCI systems are continuously improving, providing new interaction and control possibilities for both disabled individuals and healthy populations.

## 1.3 Types and Technologies of Brain–Computer Interfaces

BCI technology can be classified into various types based on signal acquisition methods, signal processing techniques, and application fields, as shown in Figure 1.4.

**1. Classification Based on Signal Acquisition Methods**
Non-invasive BCI technology uses external devices (such as electrode caps) to capture EEG signals from the scalp. This method offers advantages of safety and non-invasiveness, but its signal resolution is relatively low.

Semi-invasive BCIs collect signals by implanting electrodes on the surface of the cerebral cortex. Compared to non-invasive methods, this approach can obtain higher quality signals but requires craniotomy during the procedure.

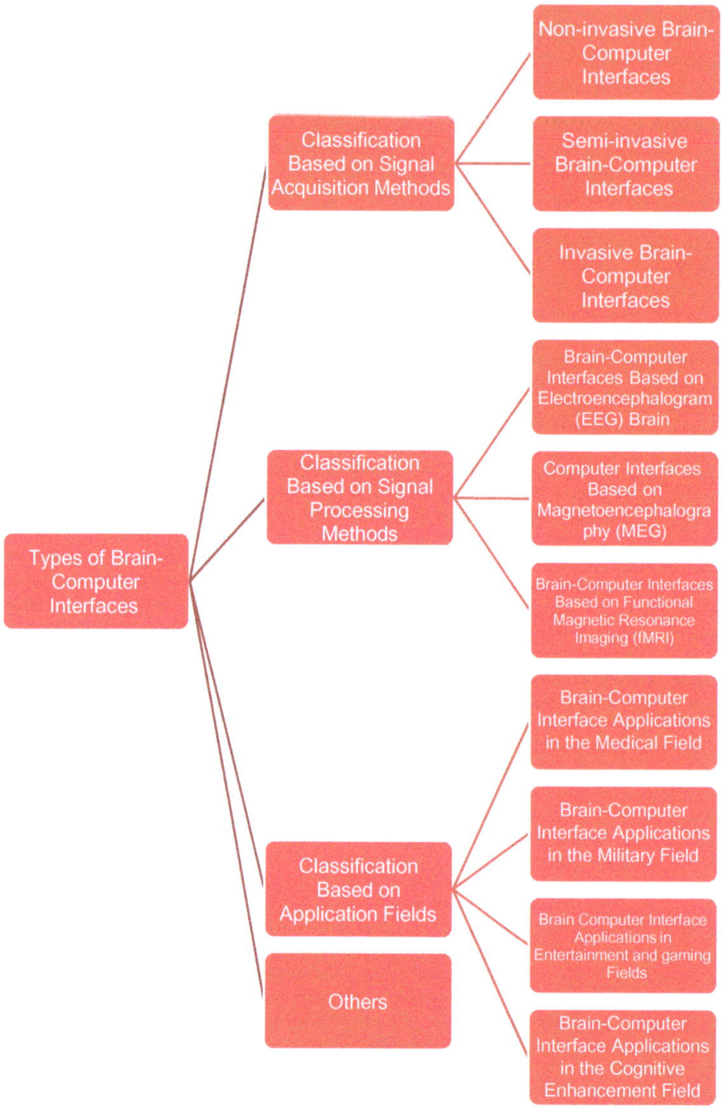


**Figure 1.4.** Types of BCIs.

Invasive BCI technology involves implanting electrodes directly into the brain, capturing signals directly from neurons or neural populations. Although this method provides the highest quality signals, it also carries the greatest surgical risks and economic costs.

**2. Classification Based on Signal Processing Methods**

BCI technology based on brainwaves (EEG signals) uses the collection of EEG signals as input to effectively control devices such as wheelchairs and computers. Additionally, MEG-based BCIs utilize magnetoencephalography technology, providing researchers with brain activity information of higher spatial resolution. Furthermore, BCIs based on fMRI use Blood Oxygen Level Dependent (BOLD) signals as a means to control external devices.

**3. Classification Based on Application Fields**

*Application of BCI in the Medical Field*: Aimed at assisting disabled individuals, especially patients with paralysis, this technology enables the control of wheelchairs or communication devices, thereby improving their autonomy and convenience in daily life.

*Application of BCI in the Military Field*: Designed to enhance soldiers' combat effectiveness, this technology enables remote control of drones or battlefield communication through BCI technology, increasing the flexibility and efficiency of combat command.

*Application of BCI in the Entertainment and Gaming Field*: Innovatively applied to game control and virtual reality experiences, providing users with a unique and immersive human–computer interaction method, enriching entertainment experiences.

*Application of BCI in the Cognitive Enhancement Field*: Focused on improving individual learning and work efficiency by real-time monitoring of attention levels, optimizing learning environments and workflows to achieve enhanced cognitive abilities.

**4. Other Technologies**

Wireless BCI technology effectively achieves wireless transmission of brain signals by utilizing wireless transmission technology, significantly enhancing users' mobility and convenience. Hybrid BCI technology integrates multiple signal acquisition and processing methods to optimize system performance and improve overall reliability. Brain-to-brain interface (B-BCI) technology further enables direct communication between brains, achieving efficient brain-to-brain information transmission.

When selecting BCI technology, it is necessary to comprehensively consider specific application requirements, cost-effectiveness, safety, and user acceptance. With continuous technological development and innovation, BCI systems are gradually exhibiting diversified characteristics, and their potential application value in various fields is continuously being explored and expanded.

**5. Future Technology Development Directions**

(1) *Wireless Transmission Technology*: To enhance user comfort and mobility, future designs of signal acquisition devices will shift toward wireless solutions. This improvement not only allows users greater freedom and convenience during use but also broadens the application scenarios of signal acquisition.

(2) *Multimodal Data Fusion Technology*: By combining various physiological signals such as EEG, fMRI, and NIRS, the stability and accuracy of BCI systems can be significantly improved. The application of this technology helps more accurately and comprehensively reflect the user's physiological state.

(3) *High-Resolution Non-Invasive Technology*: In the field of signal acquisition, future efforts will focus on developing new, non-invasive signal acquisition methods. These methods can provide signal quality close to invasive systems while avoiding the discomfort and risks that invasive methods may bring.

(4) *Efficient Algorithm Optimization*: To ensure efficient processing capabilities on resource-constrained devices, future work will further optimize signal processing and pattern recognition algorithms. These optimizations not only improve the speed and accuracy of data processing but also reduce the energy consumption and cost of devices.

## 1.4 Advantages and Limitations of Brain–Computer Interfaces

The development of BCI technology is rapid and has broad application prospects. Currently, there are several advantages.

**1. Advantages of Brain–Computer Interfaces**

(1) *Potential in the Medical Field*: BCIs show extremely broad application prospects in the medical field.

In the field of rehabilitation therapy, BCI technology can significantly assist paralyzed patients, stroke patients, and other individuals with nervous system injuries in regaining a certain degree of autonomous motor ability. By precisely decoding patients' EEG signals, this technology can control assistive devices such as prosthetics or wheelchairs, thereby improving patients' quality of life.

In terms of neural repair, BCI technology demonstrates its potential to promote neural plasticity, with the hope of helping patients gradually restore impaired brain functions, providing new approaches for the rehabilitation treatment of nervous system injuries.

Furthermore, in the field of epilepsy monitoring and treatment, implantable BCI, global devices have the capability to monitor the EEG activity of epilepsy patients in real time, enabling early detection and effective prevention of epileptic seizures, bringing hope to epilepsy patients.

(2) *Enhancing Human–Computer Interaction*: By introducing BCI technology, the efficiency of human–computer interaction can be significantly enhanced, and the quality of life for people with disabilities can be improved. This technology allows individuals with physical disabilities to precisely control electronic devices through brain signals, enabling them to perform basic operations such as typing and making phone calls in daily life, greatly increasing their autonomy and convenience. At the same time, BCI technology also offers ordinary users a new interactive experience. It breaks the traditional human–computer interaction model, allowing users to naturally and intuitively control computers and virtual reality systems through EEG signals, pioneering a more advanced and convenient way of human–computer interaction.

(3) *Applications in Entertainment and Education*: BCI technology can be widely applied in the field of video games, where the control methods it provides not only enhance the immersion of games but also offer players an intuitive and efficient operational experience. In the field of education, the introduction of BCI technology holds significant value. Through this technology, it is possible to monitor students' attention and learning states in real time, thereby tailoring personalized educational programs for each student to maximize their learning outcomes.

(4) *Scientific Research and Cognitive Science*: As a cutting-edge scientific tool, BCIs can be widely used to deeply explore human cognitive processing and fluctuations in emotional states, aiming to more accurately reveal the complex mechanisms of brain function. By closely integrating BCI technology with advanced imaging techniques (such as fMRI, i.e., functional magnetic resonance imaging), we can obtain high-precision and

detailed images of brain activity, providing strong support and evidence for neuroscience research.

(5) *Military and Aerospace Applications*: In the military and aerospace fields, BCI technology can be used to improve the combat efficiency of soldiers and pilots, or to enable communication and control in extreme environments.

**2. Limitations of Brain–Computer Interfaces**

Of course, BCIs are developing rapidly, but they also have some limitations.

(1) *Technical Challenges*

Regarding signal quality, although invasive BCI technology can provide high-quality signals, the surgical procedures involved carry higher risks. On the other hand, non-invasive brain–computer interfaces, such as electroencephalography (EEG) technology, are safer, but their signals contain more noise and have relatively lower spatial resolution.

In terms of real-time performance, achieving efficient real-time signal processing and decoding is a major challenge for current technology. Especially for non-invasive BCIs, the complex noise background further affects the speed and accuracy of signal processing.

Moreover, given the significant individual differences in each person's EEG signals, personalized calibration and training for each user are required, which undoubtedly increases the difficulty of using and promoting this technology.

(2) *Safety and Ethical Issues*

Regarding the risks of invasive surgery, the implementation of invasive BCI technology involves the surgical implantation of electrodes, a process that may carry potential risks such as infection and tissue damage. Additionally, the cost of surgery is often relatively high.

In terms of data privacy, EEG signals, as highly personalized biological data, must be collected, stored, and used in strict compliance with privacy protection measures to ensure the security and confidentiality of user information.

In addition, regarding the widespread application of BCI technology, we must carefully consider the potential ethical issues it may raise. These issues include, but are not limited to, potential violations of human privacy and possible manipulation of individual autonomous consciousness. It is necessary for us to actively explore and address these ethical challenges alongside technological development.

(3) *Application Limitations*

When facing challenges related to adaptability, BCI systems often require users to maintain a high level of focused attention. This becomes a significant limitation in practical applications, as users may find it difficult to sustain concentration for extended periods in many scenarios, thereby affecting the system's effectiveness.

Furthermore, regarding the issue of continuous training, many BCI systems require users to engage in long-term use and training, which undoubtedly increases the user burden and poses obstacles to the broader adoption of this technology.

In terms of device design, most current BCI devices still face issues of high complexity and bulky size, which cause considerable inconvenience for users in wearing and operation during practical use.

## 1.5 Potential Applications of Brain–Computer Interfaces in the Transportation Field

BCI technology has broad application prospects in the transportation field. Although research and applications in this area are still in the early stages, many potential use cases and advantages have already emerged. As shown in Figure 1.5, these are the potential applications of BCIs in transportation.

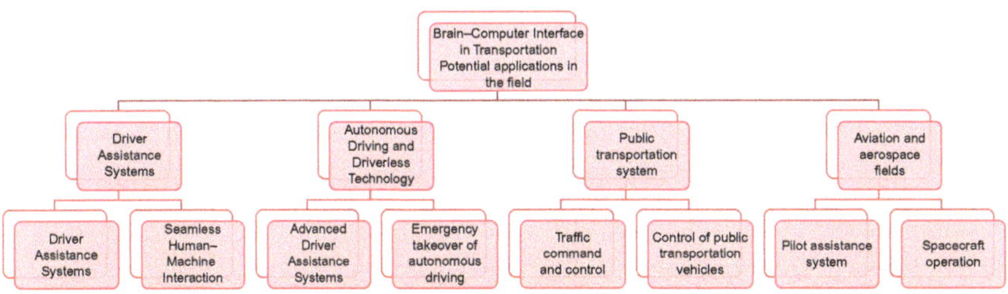


**Figure 1.5.** Potential applications of BCIs in the transportation field.

### 1. Driver Assistance Systems

(1) *Driver State Monitoring*
BCI technology has the capability to monitor drivers' EEG signals in real time, accurately assessing their levels of attention, fatigue, and emotional fluctuations. Based on the analysis of these data, BCI technology can provide immediate feedback or intervention measures, thereby effectively ensuring driving safety.

Through advanced EEG technology, we can precisely capture the driver's brainwave activity and accurately identify the driver's fatigue state. Once the system detects that the driver is fatigued, it will immediately trigger a safety alert to timely remind the driver to take necessary rest or automatically activate the autonomous driving mode to take over vehicle control, thereby minimizing potential safety risks. BCI technology can recognize the driver's emotional states, such as anger, anxiety, or excessive stress. Based on this information, the vehicle can automatically adjust the interior environment (such as music and temperature) to help the driver relax, thus enhancing driving safety.

(2) *Seamless Human–Machine Interaction*
By introducing BCI technology, a more natural and intuitive vehicle control method can be realized, significantly reducing the driver's cognitive load. In terms of vehicle function control, BCI technology allows the driver to directly operate in-car multimedia systems, navigation systems, windows, and air conditioning devices through EEG signals, greatly reducing the need for manual operations and thereby substantially improving driving safety and convenience. In some cases, BCI technology enables the driver to control certain vehicle functions directly through thought, such as steering, acceleration, or braking. This is especially useful for drivers with mobility impairments.

### 2. Autonomous Driving and Driverless Technology

(1) *Advanced Driver Assistance Systems*: BCI technology can be widely applied in Advanced Driver Assistance Systems (ADAS) to provide users with a more intelligent driving experience. Among these, the driving mode switching function is particularly prominent. Drivers can swiftly and efficiently switch between manual driving, assisted driving, and autonomous driving modes solely through EEG signal commands, without relying on

traditional button presses or voice commands, significantly enhancing driving convenience and intelligence.

(2) *Emergency Takeover in Driverless Vehicles*: When autonomous vehicles face emergency situations requiring manual intervention to ensure driving safety, BCI technology demonstrates significant advantages with its rapid and smooth control capabilities. Specifically, when the autonomous driving system detects potential danger but fails to respond in time, the BCI system can quickly recognize the driver's emergency signals, such as nervousness or panic, and immediately take over vehicle control to perform emergency braking operations, ensuring driving safety.

**3. Public Transportation Systems**

(1) *Traffic Command and Control*: BCI technology demonstrates significant application potential in the field of traffic command and management, especially in improving the command efficiency of dispatchers. Specifically, in traffic monitoring and command centers, dispatchers can utilize BCI technology to quickly select and control various surveillance cameras, traffic signals, and warning systems through EEG signals, thereby greatly enhancing response speed and command efficiency to ensure smooth and safe traffic operation.

(2) *Control of Public Transportation Vehicles*: In specific scenarios, such as for people with disabilities or particular operating environments, BCI technology demonstrates its unique assistive functions, especially in the control of public transportation vehicles, significantly improving travel efficiency and comfort. In the field of intelligent buses, the application of BCI systems enables drivers or operators to assist in controlling the bus through EEG signals during emergencies or situations requiring highly precise operations. This not only enhances driving safety but also improves the timeliness of emergency responses.

**4. Aviation and Aerospace Fields**

(1) *Pilot Assistance Systems*: BCI technology has significant application potential in pilot assistance systems, capable of enhancing flight safety and optimizing operational convenience. Specifically, pilots can use BCI technology to control certain aircraft functions, such as activating and deactivating autopilot modes and operating communication equipment,

thereby effectively reducing the pilot's operational burden and improving the efficiency of flight mission execution.

(2) *Spacecraft Operation*: In spacecraft operation, BCI technology is designed to enable remote control or assistive system management in emergency situations to ensure mission success and astronaut safety. For space station operations, BCI technology allows astronauts to perform a series of critical basic tasks in extreme environments, such as equipment inspections and experimental operations. The application of this technology can significantly improve work efficiency and reduce operational risks, thereby ensuring the stable operation of the space station and the smooth progress of scientific research missions.

BCI technology has broad potential applications in the transportation field, involving driver assistance, autonomous driving, highway and aviation traffic management, among others. From driver status monitoring to advanced driver assistance systems, and to the control of public transportation vehicles and aerospace vehicles, BCI technology can provide safer, more efficient, and more intuitive control methods. Although these applications are still in the exploration and development stages, with continuous technological advancement, BCI applications in transportation are expected to achieve major breakthroughs, enhancing the overall safety and intelligence level of transportation systems.

# Chapter 2

# Overview of Intelligent Transportation

## 2.1 Composition and Operating Principles of Intelligent Transportation Systems

Intelligent transportation systems are a complex network constructed by the integration of various elements and cutting-edge technologies. The core objective of this system is to achieve efficient and safe transportation of people and goods. Specifically, it covers the following key components.

### 2.1.1 *Integrating Infrastructure, Management, and Services: The Operational Architecture of Intelligent Transportation*

**1. Physical Infrastructure**

*Roads and bridges*: for example, highways, urban roads, rural roads, etc.

*Railway systems*: including high-speed rail, urban rail transit (such as subways and light rail), and conventional railways.

*Aviation services*: airports, runways, terminals, etc.

*Maritime facilities*: ports, docks, canals, etc.

**2. Vehicles and Transportation Means**

*Automobiles*: passenger cars, buses, trucks, etc.

*Trains*: high-speed trains, subways, freight trains, etc.

*Aircraft*: civil passenger planes, cargo planes, etc.

*Ships*: cargo ships, passenger ships, ferries, etc.

*Non-motorized vehicles*: bicycles, electric scooters, etc.; modes of transportation, including walking.

**3. Operation Management System**

*Traffic signal control*: including traffic lights, signs and markings, signal control systems, etc.

*Dispatch system*: especially for public transportation, using real-time dispatch systems to optimize and manage vehicle operations.

*Intelligent transportation system (ITS)*: providing real-time traffic information, navigation, and incident management through sensors, cameras, GPS, and IoT technologies.

**4. Support Services**

*Fuel and charging facilities*: gas stations and electric vehicle charging stations.

*Maintenance and repair*: car repair shops and inspection centers.

*Stations, ports, and airports*: used to manage passenger and cargo transfer and waiting.

**5. Others**

*Traffic regulations and enforcement*: Traffic regulations establish behavioral norms for traffic participants, such as speed limits and compliance with traffic signals. Enforcement agencies are responsible for ensuring adherence to these regulations and penalizing violations.

*Traffic education and training*: Enhancing the safety awareness and skills of traffic participants through education and training, including driver training and traffic safety education.

*Emergency services*: Including police, fire, medical, and other emergency services that provide rapid response and rescue in traffic accidents or other emergencies.

With the support of these infrastructures, basic traffic flow can be guaranteed. Currently, traffic is intelligent traffic; intelligent transportation systems achieve efficient allocation and management of traffic resources by comprehensively utilizing various technologies and facilities. Their basic operating principles include these aspects.

### 2.1.2 *The Four Pillars of Intelligent Transportation Systems: Data, Connectivity, Automation, and Optimization*

**1. Data Collection and Analysis**

This uses sensors, cameras, GPS, and other devices to collect real-time data and analyze traffic flow, road conditions, etc. In the pulse of a smart city, this flood of real-time data becomes the nerve endings of urban management, continuously providing precise nourishment for intelligent decision-making. With the rapid development of big data processing technology, we are not only satisfied with data collection but also focus on how to efficiently and accurately mine the value behind this data.

By building advanced traffic data analysis models, we can predict traffic congestion hotspots in real time and immediately push detour routes to drivers through intelligent transportation systems, effectively alleviating urban traffic pressure. At the same time, by combining historical data with weather forecasts, we can plan in advance for roads that may be affected by adverse weather and deploy emergency measures early to ensure citizens' travel safety.

Moreover, by utilizing high-definition cameras and AI image recognition technology, we can automatically identify and record traffic violations, greatly improving the efficiency and fairness of traffic enforcement. These technologies not only reduce the workload of traffic police but also encourage drivers to consciously comply with traffic rules, jointly creating a safer and more orderly traffic environment.

In terms of road maintenance, the combination of GPS, BeiDou positioning technology, and sensor data plays a significant role. By monitoring road usage and damage levels in real time, we can promptly detect and repair potential safety hazards, ensuring the integrity of road facilities and smooth traffic flow. This data-driven preventive maintenance strategy not only reduces road maintenance costs but also enhances citizens' travel experience.

**2. Information and Communication Technology**

By adopting advanced technologies such as wireless networks and satellite communications, we can achieve efficient and accurate information exchange between vehicles, infrastructure, and passengers. Intelligent transportation not only satisfies the need for instant communication between vehicles but also extends its focus to a broader interconnected domain. Through the deep integration of Internet of Things (IoT) technology, each vehicle becomes a mobile node within a smart city, capable of seamless interaction with infrastructure such as roads, bridges, and traffic signals, while also sensing and adapting in real time to subtle changes in the urban environment.

**3. Automatic Control and Automation**

This includes autonomous driving technology and automatic scheduling systems, which are used to improve operational efficiency and safety. Intelligent transportation is not limited to autonomous driving technology and automatic scheduling systems; more cutting-edge technological applications are gradually permeating every aspect of our lives, especially in enhancing operational efficiency and safety.

For example, intelligent prediction and optimization systems are widely applied. They can accurately predict traffic flow, weather changes, passenger travel habits, and other factors based on big data analysis, and dynamically adjust traffic routes, vehicle allocation, and even infrastructure maintenance plans accordingly. For instance, before holidays or large-scale events, the system can proactively identify peak crowd areas, automatically dispatch additional vehicles to reduce passenger waiting times, and simultaneously avoid traffic congestion. Moreover, for potential extreme weather conditions, the system can issue early warnings and take preventive measures to ensure the continuity and safety of public transportation services.

**4. Intelligent Optimization**

This utilizes big data and artificial intelligence technologies to optimize traffic scheduling, route planning, and so forth, thereby alleviating traffic congestion and improving work efficiency.

For example, the deep integration of IoT technology will further enhance the intelligence level of public transportation. By installing sensors and communication devices on vehicles, stations, and even road surfaces, we can obtain real-time key information such as vehicle operating status, passenger boarding and alighting, and road traffic conditions, as shown in Figure 2.1. After processing and analyzing this information through cloud computing platforms, more refined operational strategies can be generated, such as dynamically adjusting driving routes based on real-time road conditions to reduce unnecessary detours, or arranging more transport capacity in popular stations in advance based on passenger

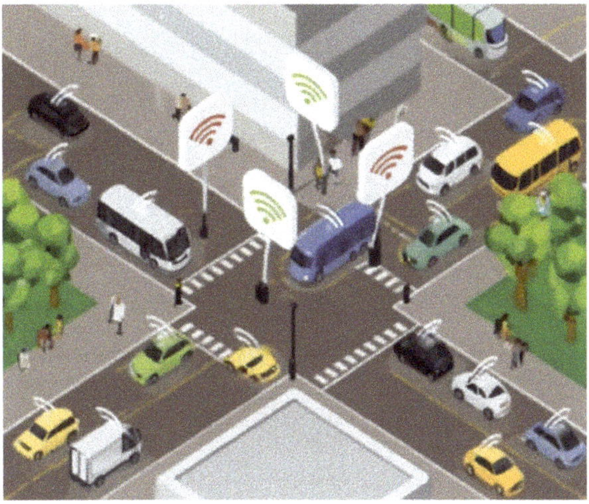

**Figure 2.1.** Real-time acquisition of traffic information.

demand forecasts. At the same time, IoT technology can also assist in achieving remote vehicle monitoring and maintenance, detecting and resolving potential faults early, thereby improving vehicle availability and operational efficiency.

### 2.1.3 *A High-Standard Ecosystem: Deconstructing the Operational Excellence of Modern High-Speed Rail*

Taking China's high-speed rail system as an example, it clearly demonstrates the components and operational principles of an intelligent transportation system. Specifically, the system consists of the following core parts:

**1. Physical Infrastructure**

*High-speed railway tracks*: Serving as the foundation for train operation, their design, construction, and maintenance all meet high standards to ensure high-speed and smooth train running.

*Stations*: Serving as passenger hubs, they not only provide comfortable waiting environments but also integrate various service functions such as ticketing, security checks, and waiting areas.

*Bridges and tunnels*: Crossing complex terrains, connecting various stations, and constituting an important part of the high-speed railway network.

**2. Transportation Vehicles**

High-speed trains, represented by "Fuxing" and "Harmony," adopt advanced technologies and design concepts to achieve high-speed, safe, and comfortable transportation services.

**3. Operation Management System**

*Train dispatching system*: relying on computer and communication technologies to realize real-time monitoring and efficient scheduling of train operations, ensuring orderly train running and smooth lines.

*Signal system*: adopting the most advanced signal control technologies to provide trains with accurate operation instructions and guarantees, ensuring train operation safety.

*Ticketing system*: integrating electronic ticketing, online booking, and automatic ticket inspection, simplifying the ticket purchasing process and enhancing passengers' travel experience.

**4. Support Services**

*In-station catering services*: providing passengers with diverse dining options to meet their dietary needs during the journey.

*Public transportation connection facilities*: setting up public transport transfer facilities at both ends of the high-speed railway to facilitate seamless connections between passengers and other transportation modes.

*Top-level maintenance services*: regularly maintaining and servicing trains and infrastructure to ensure the safety and reliability of train operations.

## 2.2 Smart Traffic Congestion and Safety Issues

In modern society, traffic congestion and safety issues are severe challenges commonly faced by major cities worldwide. These problems not only affect people's travel experience but also have negative impacts on economic efficiency and public health. Traffic congestion leads to resource waste, environmental problems, and also causes safety concerns.

**1. Traffic Congestion Problem**

What causes the traffic congestion problem? There are several reasons, as shown in Figure 2.2.

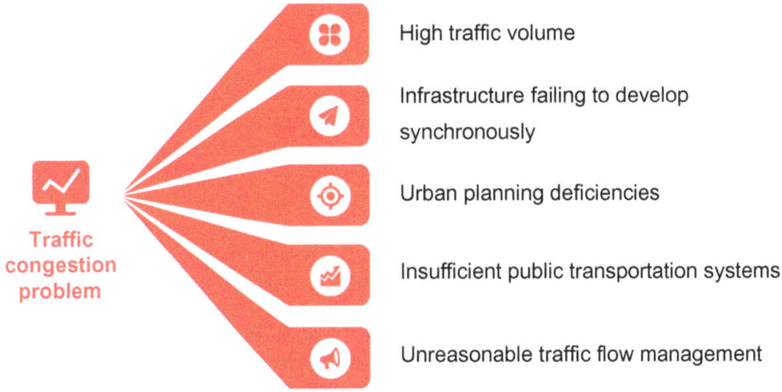


**Figure 2.2.** Traffic congestion problem.

(1) *High traffic volume*: With the rapid economic development and the sharp increase in urban population, automobiles, as a convenient means of transportation, have seen a rapid rise in ownership. This has led to a continuous increase in traffic volume on roads, even exceeding the normal carrying capacity of the roads. During certain busy periods and sections, traffic congestion has become increasingly severe, causing great inconvenience to people's travel.

(2) *Infrastructure fails to develop synchronously*: In many places, the construction of transportation infrastructure has long lagged behind the growth of vehicles and the pace of urban and rural development. The scale and quality of roads, bridges, tunnels, and other transportation facilities cannot meet the increasing traffic demand. This has led to frequent traffic congestion problems, seriously affecting people's travel efficiency and quality of life.

(3) *Urban planning defects*: During the expansion of modern cities, the division of functional zones is often unreasonable, causing traffic flow to concentrate on certain road sections during peak commuting hours. This phenomenon exacerbates traffic congestion and increases the operational pressure on urban traffic.

(4) *Insufficient public transportation system*: Although modern public transportation systems have made some progress, overall, the public transportation network is still not well-developed enough to effectively alleviate the traffic pressure caused by private cars. In some cities, the coverage and operational efficiency of public transportation still need improvement.

(5) *Unreasonable traffic flow management*: In terms of traffic management, problems such as unreasonable traffic signal settings and unclear traffic guidance signs still exist in some areas. These issues lead to traffic congestion and increase the likelihood of traffic accidents.

These causes of traffic congestion have a significant impact on our production and daily life. For example, time consumption: Drivers and passengers spend a large amount of valuable time on the road. For example, increased fuel consumption, rising operating costs, and indirect economic losses caused by traffic accidents all have a significant negative impact on the economy. For example, under traffic congestion conditions, vehicle emissions and noise pollution become more prominent, causing serious environmental pollution. Long-term traffic congestion has led to widespread anxiety and fatigue among residents, which adversely affects their quality of life.

These traffic congestion problems urgently need to be resolved, and new traffic technologies, big data technologies, artificial intelligence, and other methods are all good ways to address traffic congestion.

**2. Traffic Safety Issues**

Traffic congestion and traffic safety are key issues in traffic operation and core problems that traffic management departments must solve. Intelligent transportation systems strive to address these safety issues through big data, cloud computing, artificial intelligence, and other technologies, providing a safer traffic environment for drivers and passengers.

When we explore the causes of traffic safety issues, they generally fall into the following categories:

(1) Irregular driving behaviors, such as illegal driving, drunk driving, speeding, and fatigued driving.
(2) Unreasonable road design, such as overly narrow curves, poor visibility, and unclear signs and markings.
(3) Insufficient maintenance of vehicles, such as inherent defects or neglect in upkeep, resulting in poor vehicle performance or even loss of control.
(4) Pedestrians and non-motor vehicles not following rules, such as jaywalking, running red lights, and driving against traffic, increase the risk of traffic accidents.

(5) Weather and environmental factors, such as adverse conditions like rain, snow, and heavy fog, can cause slippery roads and poor visibility.

These traffic safety hazards cause significant losses to life and property, increase medical costs, repair expenses, and insurance payouts, and consume a large amount of public resources for traffic accident handling. The psychological state of drivers and passengers is also greatly affected.

The application of new technologies such as intelligent transportation and BCIs nowadays strives to solve these problems. For example, in 2024, Baidu will deploy a large number of autonomous taxis called "Luobo Kuaipao," which not only provide convenient travel for users but also enhance road safety. Through the exploration and application of new technologies, we can effectively alleviate traffic congestion, improve road safety, and provide the public with a more convenient, comfortable, and safe travel environment.

## 2.3 Intelligent Transportation Construction and Potential Issues

Intelligent transportation is an important approach that uses advanced information technology, communication technology, control technology, and others to optimize traffic management, improve traffic efficiency, and enhance travel experience. It encompasses a series of high-tech applications such as Intelligent Transportation Systems (ITS), Internet of Vehicles (IoV), big data analytics, and autonomous driving. The construction of intelligent transportation has made significant progress worldwide, and in China, new technologies such as IoV and autonomous driving have been rapidly implemented.

### 1. Current Status of Intelligent Transportation Construction

(1) *Intelligent transportation systems (ITS)*

*Traffic monitoring and management*: Many cities have deployed video surveillance, sensors, and intelligent control systems to monitor and manage traffic flow in real time. For example, China's "Sky Net" project achieves comprehensive monitoring through a network of cameras and sensors.

*Intelligent signal control*: Based on traffic flow at different road sections and time periods, intelligent signal systems can dynamically adjust the duration of traffic lights to improve intersection throughput. For example, Amap can now display traffic light durations in real time during navigation and provide alternative route planning for drivers during congestion.

Real-time traffic information services use mobile internet and navigation software to provide drivers with real-time traffic information, route planning, and traffic alerts. For example, Google Maps can provide real-time traffic information, route planning, and estimated arrival times to help users avoid congestion.

(2) *Vehicle-to-everything (V2X)*
Vehicle-to-Vehicle communication (V2V) enables information exchange between vehicles, warning of imminent accidents, road conditions, and more. Vehicle-to-Infrastructure communication (V2I) allows vehicles to interact with traffic infrastructure such as traffic lights and road signs to optimize traffic flow. For example, Amap uses big data analysis during driving to alert drivers when a truck is rapidly approaching from behind or when there is an accident ahead, helping ensure driver safety.

(3) *Big data analysis*
Traffic flow prediction uses big data technology to analyze historical traffic data and forecast future traffic volumes and congestion. Personalized travel services provide customized travel advice and route planning based on user behavior data.

(4) *Autonomous driving*
Various technologies from Advanced Driver Assistance Systems (ADAS) to highly autonomous driving are rapidly developing. Some cities and companies have begun testing driverless vehicles. Tesla's Autopilot system can achieve autonomous driving under specific conditions, enhancing driving safety. Baidu's autonomous taxis have already been deployed in many cities in China.

**2. Potential Issues and Challenges**

Smart transportation construction provides us with a convenient, efficient, and safe travel environment, but everything has two sides. Smart transportation also brings potential issues and challenges, as shown in Figure 2.3.

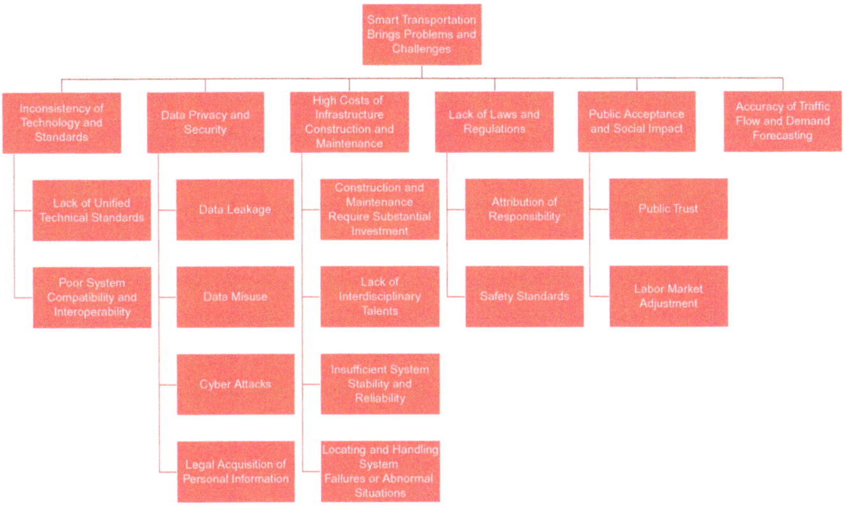


**Figure 2.3.** Issues and challenges brought by smart transportation.

(1) *Lack of uniformity in technology and standards*

Intelligent transportation involves numerous technical fields, and the lack of unified technical standards among countries and different enterprises leads to poor system compatibility and interoperability. For example, the Internet of Vehicles and autonomous driving technologies require globally unified communication protocols and security standards.

Although international organizations such as the United Nations International Telecommunication Union (ITU), the International Organization for Standardization (ISO), and the Society of Automotive Engineers (SAE) are actively promoting the formulation of cross-border technical standards and are jointly researching and developing internationally accepted standards in areas such as the Internet of Vehicles and autonomous driving, progress so far has not been rapid enough.

There are also inconsistencies in standards among enterprises. Some companies, due to competitive relationships, are not very proactive in the process of standard unification.

(2) *Data privacy and security*

The massive collection and use of data inevitably bring risks to privacy and data security. Data breaches, misuse, and cyberattacks on systems can lead to serious consequences. Driven by interests, some criminals may use technical means to illegally obtain, process, and sell personal information, severely infringing on citizens' privacy rights.

For example, in 2016, Uber suffered a serious data breach incident where hackers obtained personal information of 57 million users and drivers, including names, email addresses, and phone numbers. The compromised data included driver's license information of approximately 600,000 drivers. Uber disclosed the incident to the public only a year after it occurred and was found to have paid hackers $100,000 to delete the stolen data and conceal the event.

(3) *High infrastructure construction and maintenance costs*

The construction and maintenance of ITS require substantial investment, especially for upgrading and renovating existing roads and infrastructure. Moreover, with the deep integration of advanced technologies such as automation, big data, and cloud computing, the demand for professionals who are both technically skilled and knowledgeable in traffic management is increasing. Such interdisciplinary talents are currently scarce.

Additionally, in terms of maintenance and optimization, challenges include system stability and reliability, rapid response mechanisms, and the quick identification and handling of system failures or anomalies.

(4) *Lack of laws and regulations*

New technologies such as autonomous driving and the Internet of Vehicles involve a series of new legal issues, such as liability attribution and safety standards. Currently, these are still in the exploration and experimental stages worldwide.

(5) *Public acceptance and social impact*

New technologies need to gain public trust through extensive practice and education. At the same time, the popularization of new technologies may bring structural adjustments to the labor market, such as employment issues for traditional taxi drivers.

(6) *Accuracy of traffic flow and demand forecasting*

Although big data technology helps improve forecasting accuracy, large-scale nonlinear changes in traffic flow remain a complex problem that requires continuous optimization and improvement.

The construction of intelligent transportation is a gradual and continuously optimized process. Despite many challenges, with ongoing technological progress and gradual social adaptation, it will bring great convenience to people's travel and significant socio-economic benefits.

## 2.4 Development History of Transportation Technology

The development of transportation cannot be separated from technological support. Whether it is ancient carriages and horses or modern mechanization, transportation technology has always been closely linked with transportation itself. Transportation technology is the most important support for transportation development. With the progress of the times, new technologies and new applications are integrated into transportation, providing strong support for improving and enhancing transportation efficiency.

By reviewing the development history of modern transportation technology, we can also glimpse the origins of new technologies such as brain–computer interfaces (BCIs).

**1. Mechanization and Infrastructure Construction Stage (19th Century–Early 20th Century)**

Since the 19th century and up to the early 20th century, humanity has made significant achievements in the field of transportation technology, mainly focusing on railway transportation, automobile development, and road construction.

In the field of railway transportation, technological progress and significant improvements in application are mainly reflected in the invention of the steam locomotive and the extensive construction of railway systems. Marked by the opening of the Stockton and Darlington Railway in England in 1825, which was the world's first public railway, its inauguration signaled the beginning of a new era in railway transportation. Subsequently, the railway network rapidly expanded, a growth that not only greatly accelerated the Industrial Revolution and urbanization process but also had a profound impact on the entire socio-economic landscape.

The transformation of railway transportation greatly enhanced the efficiency of freight and passenger transport, making the circulation of goods more convenient, while also promoting the prosperity of trade and the vigorous development of the economy. This convenient mode of transportation tightened connections between regions and laid a solid foundation for subsequent economic development. Overall, the development of transportation technology from the 19th century to the early 20th century, especially the advancement of railway transportation, played a crucial role in driving social and economic progress.

In the field of automobiles and road construction, a key technological leap was the invention of the internal combustion engine, while the assembly line production model pioneered by the Ford Motor Company represented a revolutionary innovation in the automotive manufacturing industry. Exemplified by the Ford Model T launched in 1908, this production method drastically reduced manufacturing costs, transforming automobiles from luxury items into consumer goods accessible to ordinary families. This historic shift directly propelled the rapid expansion and improvement of urban roads and highway networks.

The booming automobile industry not only accelerated the historical process of urbanization but also fundamentally reshaped people's lifestyles and work patterns. It led to the birth of new urban transportation systems, greatly improving travel efficiency and convenience, while injecting strong momentum and vitality into the continuous development and prosperity of cities.

**2. Electronic and Automation Stage (Mid-20th Century)**

In the 20th century, electronic technology experienced leapfrog development, and its transformative progress profoundly changed human society while injecting powerful momentum into various fields. Automation technology also made significant strides during this period and successfully penetrated the transportation sector, bringing unprecedented convenience and safety to human travel. The aerospace field witnessed revolutionary changes, with breakthroughs in aviation and radar technology making the dream of human flight more accessible and rapid.

Specifically, after World War II, the birth of jet aircraft and the tremendous advances in radar technology played decisive roles in improving flight safety and speed. Jet aircraft, with their faster speeds and higher safety, ushered in a new era of civil aviation. During this period, aviation giants such as Boeing and Airbus began to focus more on the design and manufacture of commercial airliners. Their products not only significantly elevated the overall level of the global aviation industry but also had a profound impact on people's travel modes. This series of technological changes greatly shortened travel times between different parts of the world and provided strong technical support and guarantees for the vigorous development of international trade and the widespread exchange of cultures globally.

In terms of traffic signal control, the popularization of electronic traffic lights and the introduction of early automated traffic signal control systems symbolized a major leap in traffic management technology. For example, in the 1960s, major cities in the United States began large-scale adoption of electronic traffic lights and successfully developed preliminary traffic signal control systems. The introduction of these systems aimed to optimize urban traffic flow and improve traffic efficiency through scientific management. This initiative not only significantly enhanced urban traffic management efficiency and effectively alleviated traffic congestion but also greatly reduced the incidence of traffic accidents. The application of these technologies holds immeasurable value in improving public quality of life and ensuring road traffic safety.

**3. Informatization Stage (Late 20th Century–Early 21st Century)**
From the latter half of the 20th century to the early years of the 21st century, the informatization era entered a vibrant and dynamic golden development period. During this time, the pace of scientific and technological development was rapid and ever-changing, with many emerging technologies springing up such as mushrooms after rain and quickly permeating the transportation field. Among these, positioning technology and meteorological technology were the most prominent.

For the Global Positioning System (GPS) technology, its integration with satellite positioning technology and the official opening of civilian GPS marked the beginning of a new era. Taking the example of the United States lifting restrictions on civilian GPS use in 1995, this move directly spurred the rapid popularization of car navigation systems and GPS applications on smartphones. This transformation not only achieved precise vehicle navigation and route planning but also greatly promoted the development of the logistics industry and improved the efficiency of personal vehicle travel.

On the other hand, ITS, as a deep application of information technology in the transportation field, have also made significant progress. Taking the example of the VICS system launched in Japan in 1995, this system collects real-time traffic data and weather information to provide drivers with timely and accurate traffic information, thereby effectively optimizing traffic flow. The application of this technology not only helps alleviate traffic congestion and improve road utilization efficiency but also significantly enhances the intelligence level of traffic management.

**4. Internet and Big Data Stage (Early 21st Century–Present)**

With the development of automation and informatization, the era of the Internet has arrived. New scientific and technological advances continuously enter the transportation field for practical application. Internet, big data, cloud computing, and other technologies provide greater possibilities for transportation infrastructure construction and vehicle improvements.

The deep application of Internet technology in the transportation field has realized a close integration of technology and practical needs, especially through the construction of shared mobility platforms based on the Internet and mobile applications. Shared mobility services represented by Uber and Lyft use smartphone applications as a medium to successfully connect passengers and drivers closely, thereby disrupting the traditional taxi service model and bringing revolutionary changes to public transportation.

In the field of big data analysis, the application of technology has also demonstrated strong potential. By integrating sensor data, detailed traffic flow records, and user behavior data, we are able to conduct deep and comprehensive analyses. In this process, applications such as Google Maps, Amap, and Baidu Maps play a crucial role. They utilize users' real-time location information and rich historical traffic data to provide the public with highly accurate real-time traffic overviews and optimal route planning suggestions.

The widespread application of big data analysis technology has not only significantly improved the operational efficiency of daily transportation but also provided solid data support and theoretical basis for the long-term planning and scientific decision-making of urban transportation systems. This transformative progress undoubtedly opens new paths for the sustainable development of modern cities and the innovation of traffic management models.

**5. Internet of Things (IoT) and Intelligent Stage (21st Century–)**

Supported by core technologies such as informatization, automation, the Internet, and big data, our transportation system is gradually moving toward a new stage that is more human-centered and intelligent. With the increasingly diversified public demand for smart transportation and the background of urban agglomeration development, the requirements for transportation intelligence have reached unprecedented heights. Facing urgent issues such as traffic safety and congestion, the introduction of new technologies has become a critical problem to be solved.

Vehicle-to-Everything (V2X) technology, as a key component, enables seamless communication between vehicles (Vehicle-to-Vehicle, V2V), vehicles and infrastructure (Vehicle-to-Infrastructure, V2I), and vehicles and networks (Vehicle-to-Network, V2N). Taking Tesla's Autopilot feature as an example, this technology, through the V2X platform, achieves real-time information sharing among vehicles and interaction with infrastructure, thereby significantly optimizing the driving experience and enhancing road safety.

The widespread application of Internet of Vehicles technology has not only promoted the development of autonomous driving technology but also provided strong support for accident warning and the improvement of traffic efficiency. Looking ahead, this technology is expected to fundamentally reshape the traffic landscape and lead to profound changes in the transportation industry.

In addition, the integrated development of smart cities and ITS deeply incorporates cutting-edge technologies such as the Internet of Things, big data, cloud computing, and artificial intelligence into urban traffic management. Taking the Hangzhou smart transportation system as an example, through a network of sensors and cameras distributed throughout the city combined with big data analytics, it achieves comprehensive real-time monitoring and precise regulation of urban traffic. This system effectively improves the operational efficiency and management level of city traffic by optimizing traffic light signal timing and adjusting traffic flow distribution, greatly alleviating urban traffic congestion and providing city residents with a more convenient, safe, and comfortable travel environment.

### 6. Artificial Intelligence and Autonomous Driving Stage

In today's context, artificial intelligence and autonomous driving technologies are gradually becoming indispensable parts of our daily lives, especially in the transportation sector. The application of autonomous driving technology has become an innovative achievement that integrates AI technology, sensor technology, and real-time data processing, driving vehicles toward a high degree of automation and intelligence.

For example, Waymo's autonomous vehicles are equipped with various sensors and advanced AI algorithms, enabling them to accurately handle complex road conditions and perform efficient route planning and traffic situation assessment. Tesla's Autopilot system is also a significant milestone in autonomous driving technology, demonstrating its vast application potential. Chinese automakers such as AITO and BYD have launched multiple assisted driving vehicles, and Baidu has deployed autonomous taxis in several cities, all representing active exploration in the field of autonomous driving.

The promotion and use of autonomous driving technology are expected to bring many positive impacts. First, it is anticipated to significantly reduce the occurrence of traffic accidents and improve road safety; second, by optimizing transportation processes, it enhances transport efficiency and injects new vitality into social economic development; third, autonomous driving technology offers new travel options for those unable to drive due to physical limitations, reflecting the humanistic care of technology. Looking ahead, autonomous driving technology may fundamentally change existing private and public transportation modes, ushering in a new era of intelligent transportation.

In the field of traffic prediction and management, the combined application of machine learning and artificial intelligence technologies is gradually becoming the dominant trend. By building advanced AI algorithm models, we can achieve precise traffic flow prediction and scientific management. For instance, DeepMind, in cooperation with the UK government, uses AI technology to monitor and predict traffic flow changes in real time, enabling intelligent traffic light control and traffic flow optimization, effectively alleviating congestion. More accurate traffic forecasting not only improves the efficiency and precision of traffic management but also provides strong scientific support for urban planning, helping build smarter and more efficient transportation systems.

Overall, with continuous advancements in transportation technology, traffic systems are becoming increasingly complex and intelligent. Through the development stages and their typical application cases, we can observe the profound impact of technological progress on the socio-economy, urban development, and individual lives. In the future, as technology further advances, transportation systems will continue to transform toward greater efficiency, safety, environmental friendliness, and intelligence.

## 2.5 Sora and the Development Trends of Intelligent Transportation Technology

With ongoing technological development, we are witnessing an unprecedented transformation in the field of intelligent transportation. From the popularization of the Internet to the application of artificial intelligence and the emergence of autonomous driving technology, the transportation industry is entering a new era of digital transformation. In this new era, technological progress is profoundly changing the way we travel, making our lives more convenient and efficient.

For example, the widespread use of the Internet has greatly improved the efficiency of the transportation industry. The integration of transportation systems with the Internet has formed new transportation modes. Through ITS, we can obtain real-time traffic information, plan optimal travel routes, and avoid congestion, thereby saving time. In addition, convenient services such as online ticketing and travel reservations have also made our travel easier.

For example, the application of artificial intelligence technology has brought more innovation to the transportation field. For instance, the research and testing of autonomous vehicles are being carried out globally, which is expected to address safety hazards such as driver fatigue and drunk driving; intelligent traffic signal control systems can adjust signal timings based on real-time traffic conditions, thereby optimizing road resource allocation and improving traffic efficiency.

For example, the application of big data and IoT technologies in the transportation field is also becoming increasingly widespread. By collecting and analyzing massive amounts of data, we can better understand transportation demand and supply, providing strong support for urban traffic planning. At the same time, intelligent traffic management systems can monitor traffic conditions in real time and respond quickly to emergencies, ensuring smooth road traffic.

In addition, China is actively promoting the development of NEVs to address issues such as air pollution and energy shortages. Continuous innovation in new energy vehicle technology has improved key indicators such as the driving range and charging speed of electric vehicles, further promoting the market penetration of NEVs.

The emergence of new transportation modes such as drones and shared bicycles provides more options for short-distance travel. The application of drones in the logistics field enables fast and safe delivery of goods, reducing logistics costs. Shared bicycles offer citizens a green and convenient travel method, which helps alleviate urban congestion.

Smart transportation, empowered by technologies such as the Internet, big data, blockchain, and artificial intelligence, is full of vitality and inspires great anticipation. From Internet+Transportation to Sora+Transportation, we have experienced the convenience, safety, ease, and efficiency brought by new technologies.

**1. The Integration of Internet Technology into the Transportation Field Has Laid a New Foundation for Smart Transportation**

The popularization and development of the Internet have brought a brand-new transformation to smart transportation. In this era, Internet technology has penetrated every aspect of our lives, especially in the transportation field, where its integration has laid a new foundation for smart transportation. We can view this transformation from the following aspects:

(1) *Diversification of travel modes*: With the development of Internet technology, people's travel modes have become more diversified. Nowadays, new travel modes such as shared bicycles and shared cars have emerged, making full use of Internet technology to provide convenient and green travel options. These new travel modes have greatly alleviated urban traffic pressure, improved environmental quality, and promoted low-carbon travel.

(2) *Acquisition and transmission of real-time traffic information*: The popularization of smartphones and the development of mobile internet have enabled people to check traffic information, plan routes, book tickets, and more in real time anytime and anywhere through mobile apps. In this way, people can understand road conditions and congestion before traveling, plan the best routes in advance, greatly improving the convenience and efficiency of travel.

(3) *Big data support for traffic management departments*: Internet technology provides big data support for traffic management departments, enabling real-time monitoring and analysis of traffic information. Through big data analysis, traffic management departments can more accurately understand urban traffic conditions, providing a more scientific basis for traffic management. In addition, these data can also serve as a reference for urban traffic planning, helping optimize the allocation of traffic resources and improve road utilization.

(4) *Application of intelligent transportation systems*: The application of Internet technology in ITS is becoming increasingly widespread. ITS can realize functions such as autonomous driving and vehicle networking, further improving traffic safety, efficiency, and convenience. Autonomous driving technology is expected to reduce traffic accidents and alleviate congestion, while vehicle networking technology allows drivers to obtain real-time traffic information, avoid congested routes in advance, and improve travel efficiency. This not only enhances travel convenience but also effectively reduces fuel consumption and exhaust emissions, benefiting environmental protection and energy conservation.

**2. The Rise of ITS Leading Traffic Travel into a New Era**

Based on Internet technology, ITS have emerged. By integrating advanced information technology, communication technology, sensor technology, and more, intelligent transportation systems can monitor and predict traffic conditions in real time, providing travelers with optimal route planning and alleviating traffic congestion. Furthermore, the continuous development of autonomous driving technology is making driverless cars a reality. In the future, intelligent transportation systems will be more efficient, safer, and environmentally friendly.

China attaches great importance to the development of intelligent transportation systems, considering them an important part of strategic emerging industries. In recent years, China has continuously increased research investment in the field of intelligent transportation, achieving major breakthroughs in key technologies. On this basis, China has gradually established a complete industrial chain for intelligent transportation systems, including infrastructure, communication technology, intelligent terminals, and other segments.

The development of intelligent transportation systems is of great significance to economic and social development. First, intelligent transportation systems can improve road transport efficiency and reduce traffic congestion, thereby enhancing urban operational efficiency. Second, intelligent transportation systems can optimize resource allocation, promote the development of the logistics industry, and reduce logistics costs. Additionally, intelligent transportation systems can provide data support for government decision-making, improving the scientific and accurate management of traffic.

However, the development of intelligent transportation systems also faces many challenges. The first is technical challenges, including difficulties in system integration, data processing, and information security. The second is policy challenges, requiring the government to introduce relevant policies and regulations to regulate and guide the development of intelligent transportation systems. The third is social challenges, which involve raising public awareness and acceptance of intelligent transportation systems.

In the future, China will continue to increase investment and research efforts in intelligent transportation systems, promoting innovation and breakthroughs in key technologies. In addition, China will strengthen international cooperation, introduce advanced foreign technologies, and promote the development of intelligent transportation systems. By 2025, China's intelligent transportation system construction will achieve significant progress, providing the public with a more convenient and comfortable travel environment.

The development of intelligent transportation systems will also have a profound impact on the automotive industry. With the maturity of autonomous driving technology, driverless cars will become a reality. This will help reduce traffic accidents and improve road safety. At the same time, driverless cars will also help lower energy consumption, reduce environmental pollution, and align with the concept of green travel. In addition, driverless cars will bring new business models to the travel service industry, such as shared mobility and in-vehicle advertising.

### 3. Promotion of New Energy Vehicles to Support Low-Carbon and Environmentally Friendly Travel

As environmental pollution becomes increasingly severe, NEVs have gradually become a major trend in the transportation sector. The government actively promotes the development of the NEV industry by encouraging consumers to purchase and use NEVs through subsidies, policy support, and other measures. The NEV industry has received widespread attention and support. At the national level, the government has formulated a series of policy measures to provide strong guarantees for the development of the NEV industry. For example, policies have been introduced regarding the promotion and application of NEVs, the construction of charging infrastructure, and the recycling and reuse of power batteries, clearly defining the development direction and goals of the NEV industry. In addition, the government has increased financial subsidies and lowered production costs and consumer thresholds for NEVs, thereby promoting the rapid development of the NEV market.

In terms of technological innovation, China's NEV industry has achieved remarkable results. In recent years, domestic and foreign companies have increased their R&D investment in the NEV field, driving continuous innovation in NEV technologies. The performance, driving range, and charging speed of NEVs have all been significantly improved, making NEVs increasingly the preferred travel option for consumers.

Meanwhile, the construction of NEV charging infrastructure is also accelerating and improving. The government encourages social capital to invest in charging station construction, increasing the coverage and utilization rate of charging stations. Currently, China ranks first in the world in the number of charging stations, laying a solid foundation for the popularization and promotion of NEVs.

**4. The Rise of Shared Mobility Models, Innovating Urban Travel Methods**

In recent years, shared mobility models such as bike-sharing and car-sharing have emerged one after another, becoming a major highlight of urban travel. Shared mobility models make full use of existing resources, improve the utilization rate of transportation tools, reduce travel costs, and facilitate green travel for citizens. At the same time, shared mobility models can effectively alleviate urban congestion and promote sustainable urban development.

However, while shared mobility models bring convenience to people, they also face many challenges. First, market competition is fierce, and companies invest heavily in capital and manpower to compete for market share, which may lead to resource waste. Second, the problem of disorderly parking of shared mobility vehicles is becoming increasingly serious, affecting city appearance and traffic order. Third, safety risks cannot be ignored, such as theft of shared bikes and accidents involving shared cars. Fourth, user privacy leakage is a concern, as companies may collect users' personal information during service provision, making information security a major challenge.

In response to these challenges, shared mobility companies are also striving to find breakthroughs. On one hand, by introducing technologies such as artificial intelligence and big data, they improve the intelligence level of shared mobility services to provide users with more convenient and personalized services. On the other hand, they actively explore diversified development, such as bike-sharing companies introducing shared electric bikes and shared bicycles, and car-sharing companies trying time-sharing rentals and short- and long-term rentals to meet the travel needs of different users.

**5. Expansion of Drone Delivery Services, Opening a New Chapter in Logistics**

Drones' application in the logistics field has injected new vitality into the traditional logistics industry. Drones can deliver goods quickly and safely to their destinations, especially in remote areas and emergency situations, where they have significant advantages. With the continuous maturation of drone technology, drone delivery services will gradually expand and become a major development trend in the future logistics industry.

Drones' application in the logistics field has injected new vitality into the traditional logistics industry. Drones can deliver goods quickly and safely to their destinations, especially in remote areas and emergency situations, where they have significant advantages. With the continuous maturation of drone technology, drone delivery services will gradually expand and become a major development trend in the future logistics industry.

Drone delivery can improve logistics efficiency. In traditional logistics processes, cargo transportation is constrained by factors such as terrain and traffic, whereas drones can overcome these limitations. Drones can easily traverse complex terrains and quickly deliver goods to their destinations, significantly shortening transportation time. In addition, drones can achieve batch delivery, further enhancing logistics efficiency.

Drone delivery reduces logistics costs. Compared to traditional transportation tools such as cars and trains, drones have lower operating costs. Drones do not require fuel or drivers, and their maintenance costs are relatively low. Moreover, drones can achieve precise delivery, reducing losses during the logistics process and lowering logistics costs.

Drone delivery helps realize green logistics. Drones use clean energy and have low carbon emissions, which helps reduce environmental pollution. At the same time, drone delivery can reduce road congestion and lower the incidence of traffic accidents, thereby decreasing energy consumption and achieving the goal of green logistics.

Furthermore, drone delivery also has data collection and monitoring capabilities. During the delivery process, drones can collect logistics information in real time, providing accurate data support for logistics companies. Meanwhile, drones can monitor environmental changes during cargo transportation, such as temperature and humidity, ensuring the quality of goods.

**6. Deep Integration in the Transportation Sector to Build the Future Travel Ecosystem**

With the continuous development of technologies such as 5G and the Internet of Things, the transportation sector will achieve deep integration with other industries, constructing the future travel ecosystem. Against this backdrop, travel will become more convenient, intelligent, and personalized, allowing people to choose suitable travel modes according to their needs. Meanwhile, upstream and downstream enterprises in the industrial chain will jointly promote innovation and development in the transportation sector, achieving green, low-carbon, and efficient travel goals.

Innovation in the travel sector will also drive the development of related industrial chains. For example, the application of technologies such as intelligent transportation systems, vehicle networking, and big data analysis will foster a batch of emerging industries, providing broad development space for entrepreneurs. At the same time, travel service providers can offer one-stop travel services through cross-industry cooperation, including travel planning, booking, payment, and data analysis, enhancing user experience.

To ensure safety and stability in the transportation sector, all parties need to work together to strengthen cybersecurity defenses and prevent risks such as hacker attacks and data breaches. At the same time, laws and regulations should be improved to regulate market order and protect consumer rights.

In summary, with the continuous development of technologies such as 5G and the Internet of Things, the transportation sector will achieve deep integration with other industries, building the future travel ecosystem. In this process, governments, enterprises, and all sectors of society should join hands to promote innovation and development in the transportation sector, realize green, low-carbon, and efficient travel goals, and create a better travel life for people.

**7. Possible Applications of Sora Technology in the Transportation Sector**

In the field of transportation, Sora technology can be applied to intelligent transportation systems, intelligent vehicle networks, and other areas.

By deeply simulating the real physical world, it improves the efficiency and safety of transportation systems:

(1) *Intelligent driving assistance system*: Sora can serve as an important component of intelligent driving assistance systems by monitoring road conditions in real time, analyzing driver behavior, and predicting driving risks. It provides drivers with intelligent driving suggestions to enhance driving safety.

(2) *Intelligent traffic management system*: Sora can be applied to intelligent traffic management systems by analyzing traffic data in real time, offering traffic management departments strategies to optimize traffic signal control and improve road utilization, thereby achieving intelligent management of urban traffic.

(3) *Public transportation intelligence*: Sora can assist in realizing the intelligence of public transportation, such as predicting passenger flow, optimizing bus route planning, and monitoring bus operation status in real time, thereby improving the quality and efficiency of public transportation services.

(4) *Intelligent travel services*: Sora can provide users with personalized intelligent travel services, such as route planning, travel time prediction, and parking reservation, facilitating citizens' travel and enhancing the travel experience.

New technologies always integrate into our lives at a very rapid pace, and this integration is particularly evident in transportation. Looking back at the integration and development of intelligent transportation with the Internet, big data, cloud computing, artificial intelligence, and other technologies, we can see many opportunities and innovations, and witness how transportation becomes better, more efficient, and more convenient through innovation.

## 8. Prospects on the Impact of New Technologies such as Brain–Computer Interfaces on Future Transportation

Brain–computer interfaces (BCIs) and other emerging technologies can guide transportation toward a more intelligent, efficient, and environmentally friendly direction. By introducing advanced technologies and data analysis, traffic flow can be optimized, traffic safety improved, and congestion reduced. BCIs collect relevant information, and Sora can assist autonomous vehicles in better recognizing road conditions, predicting pedestrian behavior, and judging the driving intentions of other vehicles. Through deep learning algorithms and massive data training, Sora can analyze complex road environments in real time and provide precise driving recommendations. This will greatly enhance the safety of autonomous vehicles and reduce the incidence of traffic accidents.

In terms of autonomous driving, emerging technologies will greatly liberate humans, allowing people to move more freely; traveling from one place to another will be completed safely and efficiently by autonomous driving. However, to some extent, this may also lead to a reduction in driver positions, thereby causing employment issues. Therefore, while promoting autonomous driving technology, governments and enterprises need to increase investment in driver training and reemployment to help affected drivers transition smoothly.

BCIs invoke various information, making data security and privacy protection issues more prominent. During autonomous driving, vehicles need to continuously collect and analyze various data to ensure driving safety. This means that a large amount of personal information and sensitive data may be leaked or misused. Therefore, the improvement of relevant laws, regulations, and technologies is crucial to ensure that user data security and privacy are fully protected. Whether domestically or internationally, governments must regulate relevant participants through various laws and regulations.

In the field of public transportation, new technologies can help create more intelligent and efficient travel services. For example, subways, buses, and other public transport can use new technologies for real-time scheduling, adjusting operation plans based on passenger demand and traffic conditions. As a result, passengers' travel experience will be greatly improved, and travel times will be more accurate.

BCIs and other new technologies will also make traffic planning more reasonable and efficient. When traffic departments conduct traffic planning and construction, AI can predict traffic demand for a future period by analyzing historical and real-time data, thereby assisting governments and enterprises in formulating reasonable traffic policies. For example, in the construction of new roads and transportation hubs, AI can provide strong data support for decision-makers.

Moreover, BCIs and other emerging technologies will broaden human transportation horizons, realizing the dream of exploring the entire universe without leaving home. People will enjoy star-level travel and experience the wonders of the world.

# Chapter 3

# Overview of Neuroscience and Traffic Behavior

## 3.1 Human–Vehicle Interaction

There is a complex and multi-layered relationship between neuroscience and traffic behavior because driving behavior depends not only on individual cognitive functions and psychological states but also on various neural system mechanisms. Studying neuroscience helps understand driver behavior patterns, thereby improving traffic safety, enhancing driving experience, and optimizing traffic management.

**1. Cognitive Processes in Driving Behavior**

Driving is a complex task that involves various cognitive processes, including attention, visual processing, decision-making, memory, and motor control:

(1) *Attention*: Driving requires sustained attention, and distraction increases the risk of accidents. Neuroscience research shows that the prefrontal cortex, temporal lobe, and parietal cortex are closely involved in attention regulation during driving. Lack of attention, fatigue, or interference (such as smartphone use) can all affect driving performance and safety. Under high cognitive load, brain activity patterns also change, which has important implications for traffic management and driver training.

(2) *Visual Processing*: Visual information is the primary source of data during driving. The visual cortex and related processing areas are responsible for decoding road conditions, vehicle movements, and traffic signs. Studies

have found the effects of different lighting conditions and weather on drivers' visual perception, as well as how enhancing visual aids (such as augmented reality technology) can improve driving performance.

(3) *Decision-Making and Risk Assessment*: Driving requires rapid judgment and decision-making, involving the prefrontal cortex and related neural networks. Neuroscience research reveals how the brain balances risks and rewards in complex situations, such as lane changes in heavy traffic or crossing intersections. Understanding the decision-making process helps develop Advanced Driver Assistance Systems (ADAS) and enhances the intelligence level of autonomous driving systems.

**2. Nervous System and Driving Stress**

Driving behavior is often accompanied by stress and emotional fluctuations, all of which are influenced by the nervous system.

High-intensity traffic environments (such as peak-hour congestion) trigger stress and negative emotions, activating brain regions such as the amygdala. Prolonged high-arousal states reduce drivers' decision-making ability and reaction speed, increasing accident risk. Research combining psychology and neuroscience can explore how psychological interventions and optimizing the driving environment (such as in-car comfort) can reduce driver stress.

**3. Application of Neuroscience in Transportation Technology**

Research findings in neuroscience have begun to be applied to improve traffic management and driving experience:

(1) *Driver Monitoring Systems*: By monitoring drivers' attention, fatigue status, and emotional fluctuations through electroencephalogram (EEG) and other physiological signals, driving assistance systems can be adjusted in real time to prevent accidents. For example, EEG signals from the prefrontal cortex are used to monitor the driver's cognitive load in real time, providing timely reminders to rest or adjust driving behavior.

(2) *Cognitive Demand Optimization and Training*: Studying drivers' cognitive demands during different driving tasks can help design more effective driver training courses and examination content. Neuroscience-based simulated driving training can better enhance novice drivers' attention and reaction abilities.

(3) *Autonomous Driving and Human–Machine Interaction Optimization*: Neuroscience research can help understand human decision-making and

behavior patterns during driving, providing references for developing intelligent and human-centered autonomous driving systems. Designing human–machine interfaces (HMI) that better align with human cognitive characteristics enables drivers to interact more naturally with autonomous driving systems, such as achieving smooth transitions when handing over control.

For example, driver fatigue is a significant cause of traffic accidents. To address this issue, many modern vehicles have begun to be equipped with advanced fatigue monitoring systems.

These systems utilize EEG signal technology to monitor drivers' brainwave patterns in real time to determine whether they are in a fatigued state. As shown in Figure 3.1, several waveforms in human EEG are illustrated. When the system detects specific signal patterns closely related to fatigue in the driver's brainwaves, such as a significant increase in $\alpha$ waves (frequency range of 8–13 Hz) and $\theta$ waves (frequency range of 4–7 Hz), it immediately issues a rest reminder to the driver and, if necessary, automatically activates driver assistance functions to reduce the driver's burden.

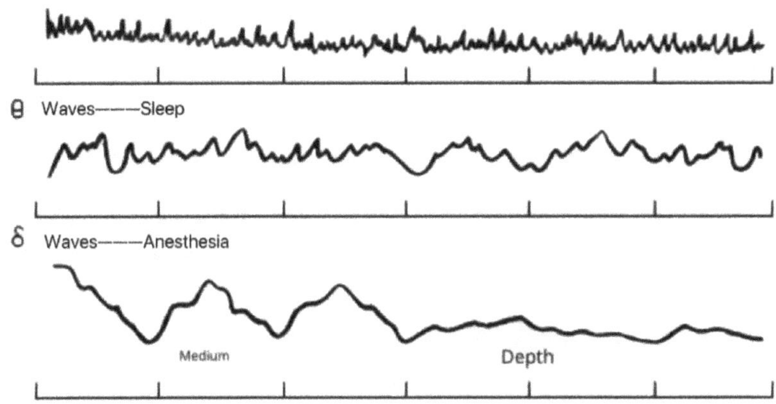


**Figure 3.1.** Several waveforms in human EEG.

It is worth noting that some high-end models have further introduced camera technology and facial expression recognition functions, combined with EEG data, to achieve more accurate fatigue detection.

The application of these technologies effectively reduces the risk of traffic accidents caused by driver fatigue and significantly improves the safety of the driving process.

This process is a very typical human–vehicle interaction scenario.

Human–vehicle interaction refers to the information exchange and communication methods between the driver and the vehicle. It covers all aspects from traditional physical interfaces such as dashboards, steering wheels, pedals, and buttons to digital and intelligent interfaces such as touchscreens, central control displays, and ADAS. Effective human–vehicle interaction can enhance driving safety, comfort, and convenience.

Taking Tesla as an example, we can see that Tesla's central control screen integrates many vehicle functions such as navigation, entertainment, air conditioning settings, and vehicle status monitoring, serving as the main human–vehicle interaction interface. We can achieve multimodal interaction through voice control, touch, and buttons on the steering wheel.

Tesla's human–vehicle interaction has many characteristics, such as an intuitive and simple interface, with a clean design of the central large screen and clear functional divisions. For example, multimodal interaction integrates touchscreen and voice interaction, providing multiple interaction methods. From navigation to vehicle control, the seamless connection of functions enhances the driving experience. Users can adjust the interface layout and function settings according to personal preferences to improve usability.

When using the autopilot function, the system continuously monitors the driver's status to ensure they can take over the vehicle at any time. In cases of suspected distraction or when the system detects that safe autonomous driving is not possible, warnings are issued with progressively escalating interventions.

Human–vehicle interaction today in intelligent driving has an increasingly wide range of applications. However, there are also some challenges.

## 4. Challenges of Human–Computer Interaction

(1) *Complexity of Human–Computer Interaction Interface Design*

For example, the issue of information overload needs to be properly addressed, with the key being to balance the richness of displayed information and the driver's actual processing capacity to ensure driving safety and prevent driver distraction caused by excessive information.

At the same time, the development of multimodal interaction technology, which integrates various interaction methods such as touch, voice, and gestures, provides new ways to enhance user experience. However, how to achieve seamless switching between these interaction methods and incorporate human-centered design concepts remains a significant challenge that needs to be addressed urgently.

(2) *User Experience and Driving Safety*
For example, regarding the real-time performance and response efficiency of human–machine interaction, the design of interaction systems must emphasize high real-time capability and low latency to ensure that drivers can respond quickly and accurately. In the interface design process, attention should be paid to its intuitiveness and ease of use, especially in driving environments where complex or cumbersome operations may seriously distract the driver, thereby increasing safety risks. Therefore, the layout and operation flow of the interface need to be carefully planned to ensure that drivers can complete various operations easily and quickly.

(3) *Personalization and Universality*
For example, considering user differences, when designing interaction methods, full consideration should be given to different users' driving habits, physical conditions (such as height and vision), and personal preferences to ensure a personalized user experience. For example, regulations and standards: Given the differences in safety standards and regulations across countries and regions, the design process must strictly comply with and meet the specific requirements of each market. For example, human–vehicle interaction design needs to consider how to use multiple human sensory channels (such as visual, auditory, and tactile) for information transmission while avoiding interference between channels.

(4) *Technology Integration and Reliability*
For example, system integration and compatibility involve the efficient integration and compatibility debugging of multiple complex systems (such as navigation, entertainment, and communication). For example, the system must be highly reliable to avoid failures at critical moments that could affect driving safety.

Research findings in neuroscience provide a scientific basis for deeply understanding the cognitive and neural mechanisms during driving, thereby contributing important technical means to improve driving safety and driving experience. This not only has direct significance for the safety of individual drivers but also has a profound positive impact on the overall optimization of traffic system efficiency and accident prevention.

## 3.2 The Influence of Attention and Decision-Making on Traffic Behavior

The influence of attention and decision-making on traffic behavior is multifaceted, relating to drivers' reaction speed, judgment accuracy, and overall driving safety.

**1. The Impact of Attention on Traffic Behavior**

Attention is the focused allocation of mental resources that helps individuals effectively extract necessary information from the environment and respond accordingly. During driving, attention must be maintained over a long period to cope with complex, dynamic, and unpredictable traffic conditions.

(1) *Attention Distraction*

Why does attention distraction occur? From external interference factors, this includes the ringing of a mobile phone, conversations from other passengers, and changes in ambient lighting. From internal interference factors, it mainly involves psychological aspects, specifically manifested as mental fatigue, emotional fluctuations, and temporary blanking of thoughts.

Multitasking situations also need to be considered, especially interactive tasks such as operating navigation systems, listening to music while driving, making phone calls, and particularly typing on a mobile phone, all of which may cause certain distractions to the driver.

The prolongation of reaction time and distraction of attention significantly increase the driver's response time to sudden situations, such as the emergency braking of the vehicle ahead or the sudden appearance of pedestrians. Moreover, this state may also lead to decision-making errors by the driver, such as violating traffic signal instructions or failing to accurately judge the distance and speed of surrounding vehicles and pedestrians.

Furthermore, when the driver is inattentive, their driving trajectory may deviate from the normal road or lane, thereby increasing the risk of traffic accidents.

(2) *Attention Management in Driving Behavior*

During driving, visual attention is crucial; therefore, visual search and fixation point management are important aspects of attention resource allocation. Drivers need to distribute their gaze among the lane, rearview mirrors, dashboard, and navigation devices. Proper attention allocation can enhance driving safety.

However, prolonged driving leads to mental fatigue and continuous decline in attention. Neuroscience research has found that sustained fatigue significantly reduces the activity of the prefrontal cortex, which is directly related to cognitive decline.

**2. The Impact of Decision-Making on Traffic Behavior**

Decision-making in traffic behavior is directly related to traffic safety. Driving requires quick and efficient decision-making abilities. For example, when changing lanes on the highway, avoiding obstacles, or responding to emergency braking, drivers must make decisions within a short time.

To make decisions in a short time, strong multi-information processing capabilities are required. For instance, speed, vehicle distance, traffic signals, and weather conditions all demand that drivers possess strong perception and information processing abilities.

So, what factors influence drivers' decision-making? As shown in Figure 3.2, there are three factors.

First is cognitive load and decision quality. When cognitive load is too high, decisions may be inaccurate or delayed. Research shows that under high pressure or complex environments, drivers tend to experience

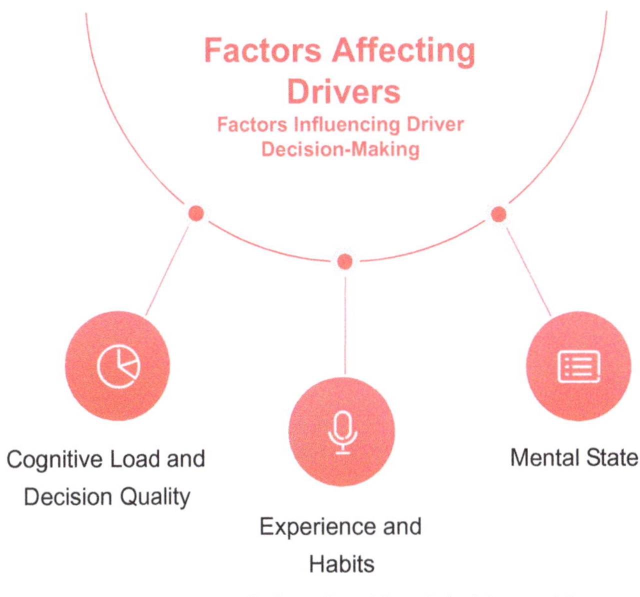


**Figure 3.2.** Factors influencing drivers' decision-making.

"tunnel vision," meaning their focus narrows and they ignore peripheral information.

Second is experience and habits. Experienced drivers usually make more accurate decisions because they can better predict and respond to complex traffic situations. However, habitual erroneous decisions may also pose threats to traffic safety.

Third is psychological state. Drivers' emotions and psychological states (such as anxiety and anger) affect the accuracy of decisions and reaction times. Negative emotions may lead to overly aggressive or overly hesitant decision-making behaviors.

**3. Interaction Between Attention and Decision-Making**

Attention and decision-making exhibit significant complementary and interactive relationships in traffic behavior. Specifically, when drivers can maintain a high level of attention and allocate it reasonably, they can quickly and accurately capture key information and make effective and appropriate decisions. Conversely, if drivers' attention is distracted or resources are insufficiently allocated, it may lead to inadequate extraction of key information, resulting in incomplete or inaccurate information upon which decisions are based. This undoubtedly greatly increases the risk of misjudgment and traffic accidents.

Excellent decision-making ability can mitigate the negative effects caused by insufficient attention in some situations. For example, experienced drivers, when facing emergencies, can still make reasonable decisions based on quick reactions and experience, even if their attention is briefly distracted.

Attention and decision-making complement each other and are inseparable in traffic behavior. By deeply understanding these two factors and how they interact, we can design safer and more efficient driving environments and assistance systems, effectively reducing the occurrence of traffic accidents. Research findings from neuroscience and cognitive psychology provide scientific foundations and technical means to improve attention and decision-making mechanisms.

## 3.3 The Relationship Between Brain–Computer Interface and Traffic Psychology

The relationship between Brain–Computer Interface (BCI) and traffic psychology offers new ideas and technical means to enhance driving safety and experience. BCI, by directly reading and interpreting brain signals, can be applied in driving for attention monitoring, emotion regulation, and driving assistance in multiple aspects.

**1. Application of Brain–Computer Interface in Traffic Psychology**

BCI can monitor the driver's attention and improve the driver's emotional state through various methods and approaches.

For example, BCI can monitor the driver's attention level in real time and detect adverse states such as distraction and fatigue. When the system detects that the driver has entered a relaxed state (high $\alpha$-wave activity) or a fatigued state (frequent $\theta$-wave occurrence), it can issue warning messages to remind the driver to rest or refocus.

By combining other sensor information (such as eye-tracking and heart rate monitoring), the BCI system can more accurately assess the driver's attention state and provide personalized interventions, such as adjusting the in-car environment (lighting, music) or activating autonomous driving mode.

This is an initial intervention. Subsequently, BCI can read signals from the driver's prefrontal cortex and other emotion-related brain regions to understand the driver's emotional state in real time, such as anxiety, anger, or fatigue. The system can identify negative emotions that may affect driving behavior and proactively take measures such as playing soothing music or adjusting the in-car atmosphere.

Additionally, the BCI system can be combined with the driver's emotion management training, using feedback mechanisms to help the driver learn how to control emotions during driving and maintain calmness and focus.

In complex driving situations (such as high-density traffic or adverse weather), BCI can assist the driver in decision-making. By directly reading brain signals, it can determine the driver's intentions (such as lane changing or accelerating) and provide decision support. The BCI system can also detect the driver's reaction time and decision quality through brain signals, switching to autonomous driving mode in a timely manner to reduce accident risks.

When the driver experiences sudden health issues (such as epileptic seizures or sudden fainting), the BCI system can quickly detect and automatically take over vehicle control to ensure the safety of the driver and passengers.

**2. Interdisciplinary Research between Neuroscience and Traffic Psychology**

Research has found that traffic psychology studies drivers' cognitive load during different driving tasks to understand its impact on decision-making efficiency and driving behavior. BCIs can monitor and assess drivers' cognitive states in real time, providing data support to help optimize driving interfaces and autonomous driving systems.

For example, by integrating BCI technology, human–machine interactions (such as augmented reality headsets) can be designed to better align with the brain's cognitive load, reducing drivers' cognitive burden and enhancing driving safety.

Additionally, neurofeedback and driver training can be conducted. Using BCI systems, drivers can improve attention, reaction speed, and emotional regulation through neurofeedback training. This training method provides real-time feedback on brainwave changes, helping individuals learn to regulate their brain functions. For instance, novice drivers can receive real-time EEG feedback during simulated driving training, gradually learning how to maintain focus in complex driving situations.

Moreover, combining BCIs with driving simulators can create a more realistic training environment, allowing drivers to undergo intensive training in a safe and controlled setting. The system can record and analyze drivers' EEG data, offering customized training programs to progressively enhance driving skills and emergency response capabilities.

The integration of BCI technology and traffic psychology offers new possibilities for improving driving safety and experience. Through real-time monitoring and intervention, it strengthens drivers' attention and emotional management, assists decision-making and automated control, and is expected to reduce traffic accidents and improve overall traffic system efficiency. Despite technical and ethical challenges, with scientific and societal advancements, the application prospects of BCIs in the traffic field will become increasingly broad.

# Chapter 4

# The Impact of Brain–Computer Interfaces on Intelligent Transportation

## 4.1 Brain–Computer Interfaces and Traffic Safety

In the current traffic system, drivers' attention, reaction speed, and physiological state are key factors affecting traffic safety. Brain–computer interface (BCI) technology can precisely evaluate drivers' current attention levels, fatigue, and psychological states by monitoring brain activity in real time, thereby providing early warnings of potential driving risks. For example, when a driver is distracted, fatigued, or emotionally unstable, the BCI system can immediately issue alerts to remind the driver to pay attention to driving safety, or even take over vehicle control through autonomous driving assistance systems to prevent accidents.

**1. Attention and Fatigue Monitoring**

Brain–computer interfaces (BCIs) can monitor the driver's electroencephalogram (EEG) activity in real time, promptly detecting the driver's attention state and fatigue level. By analyzing these data, the system can automatically issue warnings and even take certain automated measures (such as activating autonomous driving functions or adjusting the in-car environment) to enhance safety.

For example, long-haul truck drivers often face prolonged driving and nighttime driving, which easily leads to fatigue driving. A truck equipped with a BCI system can monitor the driver's EEG in real time. When the system detects a decline in the driver's attention or signs of fatigue, it can respond accordingly, such as issuing audible and visual alerts to make the driver aware that rest is needed. Through tactile feedback (such as seat vibrations), the driver is kept alert. In extreme cases, the autonomous driving mode is activated to take control of the vehicle and safely park it at a roadside rest area.

The research team from Guangdong University of Technology collaborated with Geely Automobile to conduct a study on the application of BCI technology in vehicle driving. By integrating EEG signals with the vehicle's driver assistance system, they developed an intelligent driving system capable of automatic intervention when the driver is fatigued. This technology performed well in experiments; for example, when detecting low driver attention or imminent sleep, it can issue alarms or attempt to keep the driver alert by adjusting the in-car environment (such as increasing the music volume).

**2. Automatic Response in Emergency Situations**

In sudden emergencies or when the driver loses consciousness, the BCI can immediately detect abnormalities and respond quickly. For example, at the moment of cardiac arrest or fainting of the driver, the BCI system can instantly take control of the vehicle and safely park it on the roadside, thereby preventing secondary accidents.

Suppose a driver with a history of epilepsy suddenly has a seizure while driving and loses control of the vehicle. The BCI system monitors abnormal EEG activity in real time and detects that the driver may be experiencing an abnormal condition. The system immediately switches to autonomous driving mode, stabilizes the vehicle, slows down, and searches for a suitable parking spot to safely stop the vehicle on the roadside. At the same time, the system automatically sends an alert to the medical emergency system via network connectivity, providing the location information of the vehicle and driver.

For example, Nissan has conducted multiple studies involving BCIs aimed at improving driving safety and experience. Their "Brain-to-Vehicle" technology monitors the driver's brain activity and can execute vehicle control half a second before the driver makes an operational intention. This early response can significantly enhance reaction speed and driving safety. For instance, Nissan's experimental data show that BCI technology demonstrates clear advantages in preventing rear-end collisions and improving lane-keeping accuracy.

**3. Smarter Driver Assistance Systems**

BCIs can also help improve existing driver assistance systems. Through more precise EEG monitoring, the system can better understand the driver's intentions, thereby providing smarter assistance functions during driving, such as lane keeping, automatic parking, and obstacle avoidance.

For example, in urban driving, drivers frequently need to make maneuvers such as turning, parking, and yielding. BCI system monitors the driver's EEG activity and combines it with external sensor data, such as cameras and radar, to understand the driver's intentions in real time. When the driver intends to turn or change lanes, the system makes an early judgment and assists in completing the maneuver. In complex traffic situations, the system can automatically adjust vehicle speed, steering angle, and perform real-time obstacle avoidance, reducing the likelihood of traffic accidents.

**4. Determination of Liability in Traffic Accidents**

After a traffic accident occurs, EEG data recorded by BCI can serve as a new type of evidence to help courts or insurance companies more accurately determine accident liability. These data can provide authentic information about the driver's condition, such as whether there was distraction, drunk driving, or drug-impaired driving.

Assuming two cars collide at an intersection and both drivers claim they did not violate any rules. By retrieving the EEG data recorded by the BCI system within a few minutes before and after the accident, the attention state and reaction of both drivers at the time of the accident can be analyzed. If it is found that one driver's attention significantly decreased or showed signs of fatigue before the incident, this can be used as evidence of that driver's negligence. Combined with the vehicle's dashcam data, a fairer and more scientific determination of accident responsibility can be achieved.

**5. Restoring Driving Ability**

For those who have lost their driving ability due to disability or illness, BCIs may offer a new way to drive. For example, controlling vehicle operations such as starting, accelerating, and steering through brainwaves can enable some disabled individuals to regain driving capabilities.

For instance, a driver paralyzed by a traffic accident wishes to regain the ability to drive. The BCI system monitors the driver's brainwaves through a head-mounted device and translates them into vehicle control signals. The driver controls the vehicle's start, acceleration, braking, and steering through brainwave commands to perform basic driving operations. The vehicle is equipped with automatic assistance systems that provide support when necessary to ensure driving safety. Through the BCI system, this lower-limb paralyzed driver can regain driving ability, greatly improving their quality of life and social participation.

**6. Traffic Safety Education and Training**

BCI technology can be used to simulate driving environments and assess trainees' driving skills and decision-making abilities by monitoring their EEG activity. This training method is closer to real driving experience and helps enhance drivers' safety awareness and ability to respond to emergencies.

BCI technology has broad application prospects in the field of traffic safety, but it also faces significant technical and ethical challenges. With continuous technological advancement and increased social acceptance, this field will see more development and innovation.

## 4.2 Brain–Computer Interfaces and Traffic Congestion Management

BCIs also have potential in managing traffic congestion, although research and practice in this area are still in the early stages. By integrating BCI technology with Intelligent Transportation Systems (ITS), more diversified and precise traffic congestion management solutions can be realized.

**1. Intelligent Navigation and Real-Time Decision-Making**

In many cases, the causes of traffic congestion are not limited to the inherent capacity constraints of the roads themselves but are also comprehensively influenced by multiple factors such as drivers' reaction speed, decision-making efficiency, and route selection strategies. BCI technology, as an innovative solution, can track and analyze drivers' attention levels and cognitive load states in real time. This technology, combined with detailed traffic perception data and advanced predictive models, provides drivers with highly dynamic and precise navigation guidance.

When the system detects that the driver is under high cognitive load, especially when facing complex intersections or severely congested road sections, it will automatically intervene to optimize the current driving route by selecting a smoother alternative path, thereby alleviating the driver's psychological stress and physical fatigue.

Moreover, through instant and intuitive information display and voice prompt mechanisms, the system assists drivers in more comprehensively and accurately anticipating upcoming traffic conditions, thereby encouraging them to make more scientific and reasonable route decisions.

The implementation of intelligent navigation systems not only effectively improves overall traffic flow and reduces secondary congestion caused by improper driver decisions but also significantly enhances comfort and safety during driving, making a positive contribution to reducing the incidence of traffic accidents.

**2. Traffic Management and Instant Feedback**

For urban traffic command centers, the application of BCI technology in traffic management is particularly critical. Traffic command centers are responsible for real-time monitoring and efficient dispatching of large urban traffic networks. With the aid of BCI devices, traffic managers can significantly improve their response speed to emergencies and dynamic changes in traffic flow.

The BCI system can also assist monitoring centers in accurately capturing the focus of traffic managers' attention, providing enhanced perception capabilities and decision support at critical moments. Combined with the advantages of big data analytics and artificial intelligence (AI) technologies, it can further realize precise traffic signal control and efficient execution of emergency response measures.

Therefore, BCI has broad application scenarios in improving the overall efficiency and accuracy of urban traffic management, effectively shortening traffic congestion durations, and enhancing road throughput capacity. In the event of emergencies, BCI technology can also accelerate emergency response processes, optimize resource allocation, and ensure the rapid and effective implementation of countermeasures.

**3. Shared Mobility and Optimized Scheduling**

Shared mobility services (such as Uber and Didi) play a pivotal role in urban transportation systems, and their dispatch efficiency and operational strategies have a significant impact on overall traffic conditions. To ensure efficient operation and optimize user experience, shared mobility platforms are actively exploring and applying advanced technologies, among which BCI technology is an important component.

Specifically, through BCI technology, the platform can monitor the driver's driving status and reaction speed in real time, thereby more accurately assessing their driving ability and providing a scientific basis for the formulation of vehicle dispatch strategies. On this basis, the platform combines AI models to conduct in-depth analysis of massive BCI data to predict peak passenger demand periods, and accordingly performs vehicle allocation and route planning in advance to maximize resource utilization.

During operation, the system also has the capability to adjust dispatch strategies in real time, quickly responding to market changes through intelligent algorithms, reducing vehicle empty runs caused by improper dispatch, and thereby alleviating road congestion. This not only improves the operational efficiency of shared mobility services but also significantly enhances the user travel experience.

More importantly, through more precise dispatching and route optimization, shared mobility services reduce the number of vehicles on urban roads while effectively alleviating urban traffic congestion, improving the overall traffic flow and efficiency. This series of innovative measures not only demonstrates the great potential of technology in improving urban transportation but also provides strong support for the sustainable development of future urban traffic.

**4. Regional Traffic Coordination and Diversion**

The traffic congestion problems faced by large cities are mainly concentrated in specific areas and time periods. To address this, traffic management departments can leverage BCI technology to accurately monitor drivers' behavioral patterns and psychological states under specific scenarios, thereby optimizing regional traffic flow strategies and achieving refined management.

Specific measures include dynamically adjusting traffic signal timings, flexibly implementing traffic diversion and restriction policies, effectively guiding vehicles to bypass congested areas, and reducing traffic burdens.

At the same time, intelligent signboards and navigation devices are deployed to provide drivers with optimal driving routes in real time, ensuring the timeliness and accuracy of information delivery.

BCIs have great potential in alleviating traffic pressure in key areas and improving the overall operational efficiency of traffic systems. Additionally, they help enhance drivers' adaptability and safety awareness in complex traffic environments, jointly creating a safer and smoother traffic environment.

**5. Adaptive Cruise Control and Traffic Coordination**

The Adaptive Cruise Control (ACC) system has shown significant advantages in vehicle spacing and speed control; however, its performance still has room for further optimization.

To fully tap into this potential, we propose an innovative solution that combines BCI technology with the Adaptive Cruise Control system. This solution aims to achieve precise optimization of vehicle speed and distance settings by real-time monitoring of the driver's intentions and attention focus. Specifically, the system integrates BCI data with traffic environment perception data for comprehensive analysis and dynamically adjusts the cruise speed accordingly. This strategy can effectively avoid unnecessary acceleration and deceleration operations, thereby mitigating traffic wave effects and making traffic flow smoother.

Through the above measures, we can not only significantly improve road traffic efficiency and reduce traffic congestion but also reduce fuel consumption to some extent and promote green travel. In addition, the introduction of BCI technology greatly enhances drivers' perception abilities and decision support, making the driving process more comfortable and safer.

With the development of BCI technology and the expansion of its application scenarios, its potential in traffic congestion management is promising. Through joint efforts and innovation, more practical application cases and results will gradually emerge in the future.

## 4.3 Personalized Driving Experience

BCI technology has great potential in personalized driving experiences. By reading and analyzing the driver's EEG signals, it can adapt in real time to the driver's needs and emotional states, thereby enhancing driving comfort, safety, and overall experience.

**1. Personalized Entertainment System**

During driving, selecting appropriate music and media content plays a positive role in significantly enhancing the driving experience. BCI can monitor the driver's EEG activity in real time and accurately assess their emotional states, including but not limited to relaxation, tension, and fatigue.

Based on real-time feedback of these emotional states, the system can intelligently recommend and play music or media content that matches the current emotional state. Specifically, when the system detects that the driver is fatigued, it will automatically play relaxing and pleasant music to alleviate fatigue and improve the driver's mental state; when the system recognizes that the driver is highly focused, it will select more exciting music to further stimulate their driving passion and concentration.

This not only effectively enhances the driver's emotional well-being and overall driving experience but also helps the driver maintain high alertness and concentration through the regulation function of music, fundamentally preventing safety hazards caused by fatigued driving and ensuring a safe and smooth driving process.

**2. Dynamic Cabin Environment Adjustment**

The comfort of the cabin environment, including but not limited to temperature, brightness, and seat position, plays a crucial role in enhancing the driving experience. BCI system has the ability to monitor the driver's comfort and fatigue status in real time and can automatically perform fine adjustments to the cabin environment accordingly.

Specifically, the BCI system can capture and analyze the driver's electroencephalogram signals, and intelligently adjust the seat angle, activate heating or cooling functions to maximize the driver's comfort. In addition, the system can dynamically adjust the interior lighting brightness and temperature according to the driver's mental state, creating a more pleasant driving environment.

This not only significantly improves the driver's comfort and overall driving experience during the journey but also effectively alleviates driving fatigue, thereby further enhancing driving safety.

**3. Personalized Driving Modes**

Different drivers exhibit diverse driving styles and preferences.

The BCI system can accurately analyze the driver's EEG signals, gaining deep insights into their driving tendencies, whether inclined toward stable driving or seeking an intense driving experience.

Based on these personalized analyses, the system can intelligently adjust the vehicle's power output, optimize stability control systems, and suspension settings to perfectly match the driver's unique preferences. For example, when detecting a preference for aggressive driving, the system immediately enhances power response speed, granting the vehicle more exhilarating performance; whereas if the driver prefers a comfortable driving environment, the system automatically switches to a smoother driving mode, ensuring a pleasant and relaxing journey. This not only significantly improves vehicle controllability and driving enjoyment but also profoundly satisfies the personalized needs of different drivers, comprehensively upgrading the pleasure of the driving experience.

**4. Personalized Navigation and Route Planning**

Traditional navigation systems often focus solely on algorithms for the shortest or fastest routes, neglecting the important dimension of the driver's emotional state and personalized preferences.

The BCI system injects new vitality into navigation systems. It has the capability to monitor the driver's emotions and personalized preferences, such as a liking for scenic routes or a desire to avoid complex urban traffic, all of which can be precisely captured.

On this basis, the BCI system can cleverly integrate real-time traffic information with the driver's emotional state, achieving deeply personalized route planning. Specifically, when the system senses that the driver is relaxed and has ample time, it tends to recommend picturesque roads to make the driving journey more enjoyable; conversely, if the driver is tense and pressed for time, the system quickly adjusts to the optimal fastest route, ensuring an efficient and smooth trip. This not only greatly enhances comfort and enjoyment during driving but also deeply reflects respect for and fulfillment of the driver's personalized needs, strongly promoting the development of navigation technology toward greater intelligence and humanization.

Although the application of BCI technology in personalized driving experiences is still in its early stages, some related studies and experiments have already demonstrated its tremendous potential.

For example, Nissan Motor Corporation developed the B2V technology, which analyzes the driver's brainwave signals to interpret driving intentions and adjusts vehicle parameters in advance. For instance, this technology can prepare a few milliseconds before the driver is about to turn or brake, enabling a smoother and more responsive driving experience. This technology has the potential to significantly enhance driving comfort and safety.

For example, Hyundai Motor Company has developed an emotion recognition system using BCI technology that monitors the driver's emotional state in real time by analyzing brainwave signals. This system can automatically adjust the in-car environment based on the driver's emotions, such as changing music, adjusting seats, and temperature settings, thereby providing a personalized driving experience.

In the future, with the continuous advancement and increasing maturity of BCI technology, along with the joint efforts of multiple parties, the application of BCIs in personalized driving experiences will gradually become a reality, offering drivers safer, more comfortable, and more personalized driving experiences.

## 4.4 Brain–Computer Interface and Green Mobility

BCI technology holds great potential in promoting green mobility. By real-time monitoring and interpreting users' brainwave signals, it can not only optimize the driving experience but also encourage more environmentally friendly and energy-saving travel methods.

### 1. Intelligent Driving and Energy Management

Adaptive driving modes play a significant role in optimizing vehicle energy consumption and emissions. Specifically, this mode integrates advanced BCI technology to accurately monitor the driver's intentions and emotional states, including but not limited to efficient driving, comfortable driving, and sporty driving styles. Based on this real-time data, the system can intelligently and dynamically adjust the vehicle's driving mode and energy management strategies to ensure a high degree of alignment between the driving process and the driver's current state.

Furthermore, the system possesses strong self-optimization capabilities, allowing it to finely regulate power output, energy recovery efficiency, and air conditioning system operation strategies based on the driver's immediate feedback, thereby maximizing energy utilization and minimizing energy consumption to achieve optimal energy-saving effects.

Adaptive driving modes not only significantly improve energy utilization efficiency and effectively reduce fuel consumption and exhaust emissions but also provide users with a more personalized driving experience, strongly promoting the popularization and practice of environmentally friendly travel concepts.

**2. Intelligent Navigation and Route Optimization**

The eco-driving navigation system is an innovative navigation solution designed to go beyond the traditional navigation systems' sole pursuit of the shortest or fastest route, placing greater emphasis on ecological friendliness.

The eco-driving navigation system introduces BCI technology to monitor and analyze the driver's EEG signals, gaining deep insight into their attitudes and preferences toward environmental protection, especially their preference for energy-saving routes. This unique feature enables the system to more accurately capture the driver's environmental intentions.

After comprehensively considering real-time traffic data, geographic information, and the driver's environmental preferences, the system intelligently recommends the most energy-efficient driving routes. For example, it may guide the driver to avoid heavily congested sections and instead choose roads with smoother traffic and relatively flat terrain, reducing the frequency of braking and acceleration during the vehicle's journey, thereby achieving energy saving and emission reduction.

**3. Environmental Awareness and Driving Behavior Correction**

The driver's behavior habits during driving significantly impact the vehicle's energy consumption and emission levels. To improve driving efficiency and environmental friendliness, the BCI can monitor and analyze various driving behavior patterns, including but not limited to frequent rapid acceleration and sudden braking.

By integrating real-time feedback mechanisms, including but not limited to visual alerts and auditory prompts, the BCI system promptly communicates the evaluation results of the driver's behavior. This helps the driver identify and correct poor driving habits in a timely manner, guiding them to adopt smoother and more energy-efficient driving styles. Through continuous feedback and guidance, the system not only enhances driving safety but also achieves significant long-term reductions in vehicle fuel consumption and emissions.

Driving behavior feedback helps reduce vehicle fuel consumption and emissions over the long term, contributing to environmental protection; on the other hand, it effectively extends the vehicle's service life and lowers maintenance costs for the owner. More importantly, the system plays an irreplaceable role in cultivating drivers' eco-driving habits and enhancing their awareness of green travel.

**4. Integration with Shared Mobility Platforms**

Shared mobility, as an important approach to promoting green travel, has shown remarkable results; however, the current usage efficiency and user experience of shared vehicles still require further improvement and optimization.

BCI technology deeply integrates into shared mobility platforms, enabling more precise and personalized services by monitoring users' emotional states and specific travel needs. It can capture and analyze users' physiological and psychological conditions in real time, combining their historical preferences and current demands to dynamically adjust and optimize the allocation strategy of shared vehicle resources.

For example, when a user feels fatigued or unwell, the system can quickly respond by automatically dispatching the nearest electric shared vehicle and planning the most reasonable driving route for the user based on traffic conditions and energy-saving standards. This not only effectively improves the utilization efficiency of shared vehicles but also greatly enhances the user's travel experience and satisfaction.

In the long run, the integration of intelligent shared mobility platforms with BCIs will strongly promote the widespread adoption and efficient utilization of shared electric vehicles, thereby reducing the overall carbon emissions in the transportation sector and making a significant contribution to achieving green and low-carbon travel methods.

With continuous technological advancements and the expansion of application scenarios, BCI technology shows tremendous potential in promoting green travel. It can not only personalize and optimize the driving experience but also facilitate more environmentally friendly and energy-saving travel methods through intelligent energy management, route optimization, and driving behavior correction.

## 4.5 Brain–Computer Interfaces and Government Policy Supply

General Secretary Xi Jinping emphasized that "Artificial Intelligence (AI) is a strategic technology leading this round of scientific and technological revolution and industrial transformation, possessing a strong 'leading goose' effect with significant spillover and driving capabilities." The State Council's 2017 "New Generation Artificial Intelligence Development Plan" clearly states that "by 2030, the overall level of AI theory, technology, and applications will reach a world-leading level, making China a major global AI innovation center." To achieve this goal, China has comprehensively increased investment in AI, promoting profound changes in multiple fields such as intelligent manufacturing, intelligent transportation, and urban management. Currently, AI has become a key driving force for technological innovation and economic transformation, and BCIs are a critical element in the development of AI.

With the rapid development of technology, AI has gradually become an important force driving social progress. However, during the rapid growth of AI, its potential security risks have also become increasingly prominent. Especially for technologies such as BCIs, which on one hand collect various human information and on the other hand provide feedback related to that information, this not only involves personal privacy but also concerns personal safety and human ethics.

These risks not only restrict the further development of AI technology but also have negative impacts on the political, economic, and social security of China and the world. Therefore, how to simultaneously advance AI technology and enhance AI security governance capabilities has become a global consensus and an urgent task.

At the technical level, the security risks of BCIs mainly include data security, algorithm security, and system security. As large amounts of data are used to train AI models, data security issues have become increasingly prominent. On one hand, data privacy protection has become a key topic, with the security of corporate and personal privacy information under threat; on the other hand, issues such as data discrimination and data misuse may lead to unfairness and social conflicts. Additionally, the opacity, lack of explainability, and potential biases of AI algorithms make algorithm security an issue that cannot be ignored. Unreasonable or unfair algorithms may lead to unjust decisions, thereby affecting social stability. At the same time, the security of AI systems is also crucial. If systems are hacked or maliciously exploited, it could have serious consequences for national security, public safety, and other areas.

At the political level, the AI security risks influenced by BCIs will impact national political stability. On one hand, hostile forces may use AI technology for cyberattacks, espionage, and public opinion manipulation, posing threats to China's political security. On the other hand, globally, the development and application of AI technology may lead to new digital divides and political imbalances. The gap between developed and developing countries in the AI field will further intensify international political competition and even trigger geopolitical tensions.

In the economic domain, the AI security risks influenced by BCIs are mainly reflected in finance, energy, infrastructure, and other areas. The widespread application of AI technology has made critical infrastructure more intelligent and interconnected but has also increased the risk of cyberattacks. Once critical infrastructure is damaged, it may lead to national economic recession or even social unrest. Moreover, the application of AI technology in the financial sector also carries risks. For example, errors made by AI assistants in financial investment and risk assessment may cause fluctuations and instability in financial markets.

At the social level, the AI security risks influenced by BCIs will negatively affect the quality of life and mental health of the public. As AI technology is applied in education, healthcare, employment, and other fields, people increasingly rely on intelligent devices and services. However, if AI systems malfunction or are maliciously exploited, it may cause disruptions in people's lives, strain medical resources, and lead to unemployment. Furthermore, excessive dependence on AI technology may increase psychological stress and weaken interpersonal communication skills, thereby affecting social harmony.

General Secretary Xi Jinping emphasized during the ninth collective study session of the 19th Central Political Bureau of the CPC Central Committee that "it is necessary to strengthen the assessment and prevention of potential risks in AI development, safeguard the interests of the people and national security, and ensure that AI is safe, reliable, and controllable." AI security has become a key link in technological innovation and social development. Therefore, we need to effectively address the challenges of AI security risks, accelerate the improvement of AI security governance capabilities, ensure the healthy, orderly, and safe development of AI, and continuously promote the digital and intelligent transformation and upgrading of industries.

The government encourages the development of BCI and other AI technologies on one hand, and on the other hand, standardizes their development by providing policy support and establishing corresponding regulatory and protective measures through laws and regulations. This not only ensures the development of new technologies such as BCIs but also provides binding constraints for their healthy development to safeguard the interests of the country and the people.

**1. The legal and regulatory system for artificial intelligence technologies such as BCIs will be introduced and improved**

The widespread application of BCI technology may lead to issues such as privacy breaches, security risks, and social ethics. For example, in the medical field, the security and privacy protection of patient data is crucial. In addition, with technological advancement, the risk of hackers attacking BCI devices is also increasing. From an ethical perspective, BCI technology may provoke reflections on human identity and consciousness.

To address the challenges and risks brought by BCI technology, the Chinese government will gradually introduce relevant laws and regulations to ensure that the technology develops in a compliant and safe manner. The possible directions for laws and regulations are as follows:

(1) *Data security and privacy protection*: Regulate the collection, storage, processing, and transmission of patient data by BCI devices, and increase penalties for leaking patient information.
(2) *Product quality and safety supervision*: Formulate production, sales, and usage standards for BCI devices to ensure product quality and prevent unqualified products from entering the market.
(3) *Clinical trials and ethical review*: Strengthen the supervision of clinical trials of BCI technology to ensure that the trial process is legal and compliant, fully considering the rights and interests of patients.
(4) *AI and human collaboration*: Clarify the responsibilities and authorities of AI in fields such as healthcare and education, regulate human–machine collaboration models, and ensure human dominance in the decision-making process.
(5) *Social ethics and moral norms*: Guide the research and application of BCI technology to follow ethical principles, preventing technological abuse and unethical behavior.

**2. Provide policy support to promote fairness, transparency, and explainability of AI algorithms, reducing potential bias and discrimination**

(1) *Policy level*: The Chinese government attaches great importance to the fairness, transparency, and explainability of AI algorithms. Relevant departments will issue a series of policies and regulations clearly specifying requirements for AI applications in data collection, algorithm design, and technology development. For example, in AI applications in finance, healthcare, and other fields, enterprises are required to follow principles such as minimization, purpose limitation, and consent during data collection and processing to ensure data security and protect personal privacy. In addition, AI algorithms must fully consider fairness, transparency, and explainability during the design process to reduce potential bias and discrimination.

(2) *Technical research and development*: Enterprises and research institutions will be encouraged to increase investment in developing AI algorithms with independent intellectual property rights. Through technological innovation, the accuracy, reliability, and security of algorithms will be improved, thereby reducing potential bias and discrimination. At the same time, enterprises are encouraged to adopt explainable AI technologies to make algorithms more transparent in the decision-making process, helping eliminate public doubts and concerns about AI.

(3) *Promote AI talent cultivation*: Provide policy support to improve the overall industry level. By strengthening the construction of disciplines in the AI field, establishing related majors, and increasing efforts in talent cultivation, a group of AI professionals with both expertise and ethical awareness will be trained. These talents will become an important force in promoting fairness, transparency, and interpretability of AI algorithms, providing solid support for the development of China's AI industry.

(4) *Popularization aspect*: Efforts to promote AI ethics and related laws and regulations will be intensified to raise society-wide awareness of algorithmic fairness, transparency, and explainability. Through various publicity activities and training courses, the general public will be informed about AI ethical principles, guiding enterprises, researchers, and the public to jointly focus on issues of fairness, transparency, and explainability in AI algorithms, thereby fostering a positive social atmosphere.

**3. Enact relevant laws and regulations to enhance AI system security, strengthen cybersecurity defenses, and prevent hacking and malicious exploitation**

China will further improve relevant laws and regulations to clarify the legal responsibilities for hacking AI systems. Illegal activities will be severely punished according to the law to effectively protect the security of AI systems such as BCIs. In addition, supervision over AI systems in data processing, transmission, and other aspects will be strengthened to ensure data security.

Moreover, in terms of international cooperation, China will actively participate in international cybersecurity affairs and strengthen exchanges and cooperation with other countries in the security field of AI systems, such as BCIs. Together, they will address global cybersecurity challenges and maintain peace and stability in cyberspace.

# Part II

# Brain–Computer Interface (BCI) in Transportation

# Chapter 5

# Brain–Computer Interfaces (BCIs) and Traffic Control

## 5.1 Brain–Computer Interfaces (BCIs) and Traffic Signal Control

Nowadays, brain–computer interface (BCI) technology is rapidly emerging as an innovative force in the field of traffic management. By directly converting human brain signals into computer commands, BCI technology brings a brand-new intelligent solution to traffic signal control. In the future, this technology is expected to significantly improve the efficiency and safety of traffic flow management, fundamentally changing the traditional traffic control model.

As a key component of traffic management systems, the efficiency and responsiveness of traffic signals directly affect the capacity and safety of the entire traffic network. Traditional traffic signal control relies on preset programs and manual intervention, making it difficult to flexibly respond to real-time traffic conditions. BCI technology, by real-time collection and analysis of EEG signals from traffic managers or drivers, can instantly adjust the duration and switching frequency of traffic lights, thus achieving more precise and dynamic control. This intelligent control method not only helps alleviate traffic congestion but also significantly reduces traffic accidents caused by human operational errors.

The application prospects of BCI technology in traffic signal control are very broad. With the continuous advancement of neuroscience, artificial intelligence, and big data technologies, the accuracy and response speed of BCI systems will be further improved, and their application scope will continue to expand. The deep integration of BCI technology with other intelligent transportation systems, such as autonomous driving and the Internet of Vehicles, will build a more intelligent and collaborative traffic management network. The coordinated operation of multiple systems can not only enhance overall traffic management efficiency but also provide more personalized and humanized travel services.[1]

The application of BCI technology in traffic signal control will also promote the development of intelligent transportation and enhance the overall intelligence level of traffic systems. By real-time monitoring of traffic flow, weather conditions, and emergencies, BCI systems can dynamically adjust signal control strategies to optimize traffic flow, reduce vehicle waiting times, and improve road capacity. At the same time, by monitoring drivers' fatigue and attention states, BCI systems can provide early warnings and take intervention measures, further improving traffic safety.

However, the application of BCI technology also faces many challenges. Technical issues such as the accuracy of signal acquisition and processing, system stability, and response speed still need to be further addressed. At the same time, legal and ethical issues such as data privacy protection and technology regulation require attention and resolution. Nevertheless, the enormous potential and development prospects of BCI technology in traffic management cannot be ignored.

Next, we delve into the specific applications of BCI technology in traffic signal control across highways, water transport, railways, and civil aviation. Through the development of these different transportation sectors, as well as the practical operation methods and application effects of BCI technology in these fields, we comprehensively understand how BCI technology exerts its unique advantages and roles in intelligent traffic management.

### 5.1.1 *Brain–Computer Interface and Highway Traffic Signal Control*

With the rapid development of technology, BCI technology has gradually become an important innovative force in the field of traffic management.

[1] The Internet of Vehicles primarily refers to the effective utilization of dynamic information from all vehicles within the information network platform through wireless communication technology by onboard devices in vehicles, providing various functional services during vehicle operation.

By directly reading and interpreting brain signals, BCI technology provides an intelligent and humanized solution for highway traffic signal control. This technology not only improves the efficiency of traffic flow management but also significantly enhances traffic safety and response speed.

In the application of BCI to highway traffic signal control, the complexity and real-time nature of traffic management must be fully considered. The core equipment includes electroencephalogram (EEG) acquisition devices, signal processing units, and traffic signal controllers. The EEG acquisition devices use non-invasive head-mounted devices to capture the brain signals of traffic managers or drivers in real time, and these signals are then transmitted to the signal processing unit for processing and analysis.[2]

The implementation steps of BCI in the field of highway traffic signal control mainly include three steps, as shown in Figure 5.1. First, install EEG acquisition devices and signal processing units at traffic management centers or key intersections; second, ensure these devices can stably acquire and transmit EEG signals; finally, the signal processing unit analyzes the signals and executes these instructions through the traffic signal controller.

The traffic signal control logic is the core of the BCI system, determining how the processed EEG signals are converted into specific traffic signal control commands. Advanced control algorithms can not only recognize the intentions of traffic managers but also dynamically adjust the switching frequency and duration of traffic signals based on real-time traffic conditions.

For example, during peak hours, the system can automatically extend the green light duration based on real-time monitoring data to reduce vehicle congestion; in the event of traffic accidents or emergencies, the

---

[2]EEG is a method that uses electrophysiological indicators to record brain activity. It captures the collective postsynaptic potentials generated by the synchronized activity of large numbers of neurons during brain function. EEG records fluctuations in electrical waves during brain activity, representing the overall reflection of the electrophysiological activity of brain nerve cells at the cortical or scalp surface. These brain waves originate from the postsynaptic potentials at the apical dendrites of pyramidal cells. The formation of synchronized rhythmic patterns in brain waves is also associated with the activity of the nonspecific thalamocortical projection system. As fundamental research in neuroscience, EEG monitoring is widely applied in clinical practice.

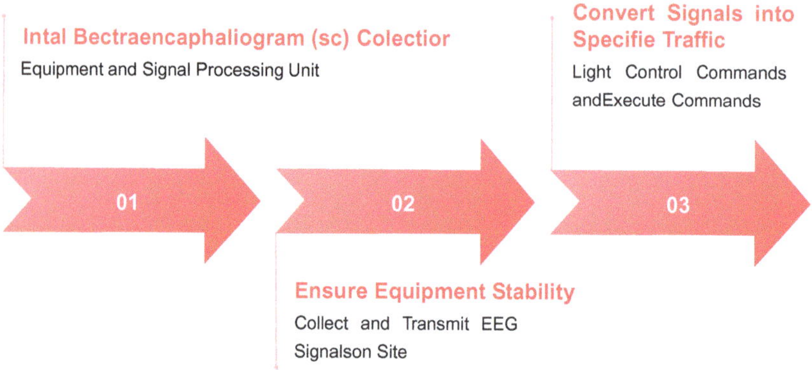


**Figure 5.1.** Implementation steps of brain–computer interface in highway traffic signal control.

system can quickly adjust the traffic signal status to provide priority passage for emergency vehicles such as ambulances. This dynamic adjustment not only improves the efficiency of traffic flow management but also significantly enhances traffic safety.

Portable BCI Driving Safety Intelligent Prevention and Control System.

In recent years, the application scenarios of BCI technology have been continuously expanding. In the Shanxi Transformation and Comprehensive Reform Demonstration Zone, whenever drivers of heavy vehicles such as dump trucks and freight trucks exhibit behaviors such as fatigue or distracted driving, the in-vehicle equipment issues a warning to minimize the occurrence of traffic accidents. This traffic signal control device was developed by Shanxi Diyitong'an Technology Co., Ltd. (referred to as "Diyitech") and is named the "Portable BCI Driving Safety Intelligent Prevention and Control System."

Diyitech utilizes portable acquisition of multiple physiological electrical signals such as EEG, electrocardiogram (ECG), electromyogram (EMG), and blood oxygen. By employing biosensors, portable wearable devices, and signal processing technology, it collects multi-dimensional biological signals in real time and analyzes them through artificial intelligence technology to digitize the driver's physiological and psychological states, providing an intelligent safety monitoring solution.

The system consists of cameras and speed radars inside the vehicle cabin, a portable EEG cap and smart wristband worn by the driver, and a backend monitoring and early warning platform. Since June 2020, the system has been successively installed on 866 heavy vehicles across 48 enterprises in the Shanxi Transformation and Comprehensive Reform Demonstration Zone, establishing a comprehensive driving safety early warning system that includes pre-warning reminders, timely alarms during incidents, and evidence-based follow-up. After triggering a warning, the system automatically records video, captures photos, uploads them to the supervision platform, and promptly reminds the driver to pay attention to safety.

During the application process, Diyitech continuously optimizes product functions, improves the service system, and enhances the system's practicality and reliability. In 2023, Diyitech participated in the Ministry of Public Security's key laboratory project "Research on the Judgment, Early Warning, and Intervention Technology of Motor Vehicle Driver Fatigue Driving Violations" and contributed to drafting and revising the public safety industry standard "Rules for Identifying Motor Vehicle Driver Fatigue Driving." Diyitech also stated that it plans to further expand the application of BCI technology in the medical rehabilitation and health fields in the future, forming a comprehensive intelligent safety monitoring solution.

The application results of the "Portable BCI Driving Safety Intelligent Prevention and Control System" demonstrate the potential of BCI technology in highway traffic signal control and also show its practical effects in improving traffic safety and reducing accident rates. Through such innovative applications, BCI technology is bringing revolutionary changes to intelligent traffic management.

The application of BCI technology in highway traffic signal control reveals its enormous potential and broad prospects. By real-time acquisition and processing of EEG signals, the BCI system can achieve intelligent and humanized traffic signal control, not only improving traffic management efficiency but also significantly enhancing traffic safety. The successful application of this technology marks the entry of traffic management into a new intelligent era and provides strong technical support for the future development of traffic systems.

### 5.1.2 *Brain–Computer Interface and Waterway Traffic Signal Control*

Under the guidance of the strategy of building a strong scientific nation, various emerging technologies are increasingly integrated with all fields of social development. As a fundamental pillar of a strong nation, transportation is naturally a top priority, and the application scope of BCIs is continuously broadening. As an important component of transportation, waterway transport faces the challenge of improving management efficiency and safety through intelligent means, which has become a targeted problem for BCI technology to solve.

Waterway transport is a crucial mode of global trade flow, with its management efficiency and safety always being focal points. Traditional waterway traffic signal control mainly relies on preset programs and manual intervention. Facing the complex and ever-changing waterway environment, it suffers from slow response speed and poor flexibility. The introduction of BCI technology provides a brand-new solution for waterway traffic signal control, enabling dynamic adjustments based on real-time EEG signals to improve management efficiency and safety.

The application logic of BCI systems in waterway traffic signal control is similar to that of highway traffic signal control. Both involve installing signal acquisition devices on key transportation tools such as ports and ships. The received signals are then uploaded to an information processing center, which converts them into control commands. The signal light controller adjusts the status of the traffic lights according to the control commands from the processing unit.

In the practical application of waterway traffic, EEG acquisition devices on ships, ports, and other locations capture the operator's brain activity through electrodes and upload the data in real time. The intelligent traffic signal processing unit uses machine learning algorithms and neural network models to process and analyze the signals, identify the operator's intentions, and generate corresponding traffic signal control commands. These commands are executed by controllers to adjust the status of the traffic lights.

Currently, although there are no widely publicized practical application cases worldwide, the industry unanimously believes that the potential application directions and prospects of BCI technology in waterway traffic signal control are very broad. These application directions mainly focus on the following aspects.

**1. Dynamic Channel Management:** BCI technology can achieve real-time monitoring of ship operators' status and dynamically adjust channel traffic lights based on the operator's attention and fatigue levels to optimize traffic efficiency. Especially during peak periods, extending green light duration can reduce congestion and improve channel throughput.

**2. Emergency Handling:** In sudden emergency situations, the BCI system can quickly recognize the operator's stress and emergency responses, prioritizing passage rights for rescue vessels to ensure efficient and timely emergency response.

**3. Integration of Automation and Intelligence:** The combination of BCI technology with autonomous ships and intelligent navigation systems can realize fully automated waterway management. For example, in remote control of unmanned vessels, precise remote operation and command can be achieved through the operator's EEG signals.

**4. Safety Monitoring and Early Warning:** BCI technology can monitor the operator's health and status in real time. When operator fatigue or distraction is detected, the system automatically issues a warning and adjusts the traffic light status to prevent safety accidents caused by operational errors.

**5. Environmental Monitoring and Protection:** BCI systems can also integrate environmental monitoring functions, such as real-time monitoring of water quality and meteorological conditions, and automatically adjusting channel signal lights to ensure the safety of shipping activities and environmental protection.

With the continuous advancement of BCI technology, the system's performance and stability will be further improved. In the future, BCI systems are expected to deeply integrate with other intelligent transportation systems, such as ship navigation systems and autonomous vessels, to build a more intelligent and automated waterway management network.

In summary, the application of BCI technology in waterway signal light control provides a brand-new solution for modern waterway management. By real-time collection and processing of EEG signals, BCI technology can realize intelligent and humanized control of waterway transportation, promptly improving the efficiency and safety of waterway management.

### 5.1.3 *Brain–Computer Interface and Railway Signal Light Control*

In the field of railway traffic management, the introduction of BCI technology heralds a future where railway signal light control will be more intelligent and efficient. The application of BCI technology can significantly improve the response speed and control accuracy of railway signal lights, thereby enhancing the safety and efficiency of railway transportation.

The railway signal light control system plays a key role in railway transportation. Traditional railway signal light control relies on complex hardware systems and manual operations. With the advent of the digital era, the application of cutting-edge technologies such as big data and AI has greatly enhanced the intelligence of railway management systems, and the application of BCI technology will further promote the intelligent upgrade of railway development.[3]

From a technical perspective, previous technologies such as big data and AI mainly focused on enhancing the intelligence of railway traffic hardware facilities, whereas BCI technology extends this intelligence enhancement to railway operators by directly extracting intentions from operators' EEG signals to perform real-time signal light control, thus addressing the issues of delay and insufficient response in traditional systems.[4]

Currently, countries worldwide are intensifying application testing of BCI technology in the railway transportation field, with the main focus on monitoring brain activity of relevant personnel to measure and monitor mental state and mental health, while analyzing concentration and emotional states to enable computers to better understand and cooperate with human emotions. The greatest achievement so far has been the analysis and detection of fatigue levels and attention of relevant

---

[3]Big data refers to datasets of such immense scale and complexity that they cannot be effectively captured, managed, processed, or analyzed using conventional software tools within a reasonable timeframe yet hold the potential to be transformed into actionable insights for informed and proactive business decision-making.

[4]Artificial Intelligence (AI) is a key driver of the new technological and industrial revolution. As a field of technical science, it focuses on developing systems that simulate and extend human intelligence. AI aims to understand intelligence and create machines capable of human-like responses, covering areas such as robotics, speech and image recognition, natural language processing, machine learning, and computer vision.

staff, preventing railway traffic hazards caused by excessive fatigue or mental burden.

In some railway research projects, multiple laboratories worldwide have explored the application of EEG signals in automated control. For example, some experiments have used EEG signals to control the operation of model trains, demonstrating the potential of BCI technology in high-precision control. These experiments show that BCI technology can provide real-time feedback in high-load operational environments, thereby optimizing system response times.

Some intelligent railway monitoring systems have already begun adopting similar technologies for status monitoring and early warning. These systems improve safety by monitoring the physiological and psychological states of operators. Although they do not directly control the signal lights, they lay the foundation for the application of BCI technology in the railway sector.

Relevant information shows that currently Nanyang Technological University in Singapore, Samsung in South Korea, the Singapore Institute of Infocomm Research, University of Technology Sydney in Australia, Beijing Institute of Technology, Shanghai Jiao Tong University, Beihang University, Northeastern University, Beijing University of Technology, Harbin Institute of Technology, and Huazhong University of Science and Technology are conducting research and validation based on non-invasive BCIs and machine learning. Analysis of these studies indicates that the application scenarios and development directions of BCI technology in railway transportation mainly focus on three aspects, as shown in Figure 5.2.

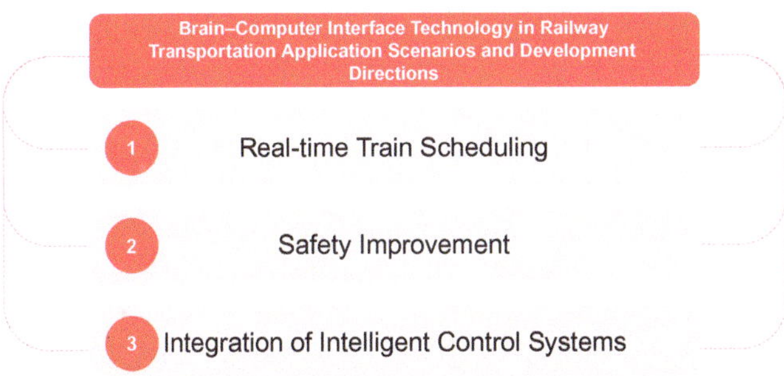


**Figure 5.2.** Application scenarios and development directions of brain–computer interface technology in railway transportation.

**1. Real-time Train Scheduling:** BCI technology can achieve real-time monitoring of train controllers' status and automatically adjust signal light states to optimize train scheduling. For example, when the operator's attention level decreases, the system can automatically make adjustments to prevent scheduling errors caused by human factors.

**2. Safety Enhancement:** The BCI system can monitor the operator's fatigue level and emotional state, and adjust signal light control strategies accordingly. For instance, when the operator is fatigued, the system can increase the frequency of warning signals and reduce train speed to enhance operational safety.

**3. Integration of Intelligent Control Systems:** With technological development, BCI technology can be integrated with automated train control systems and intelligent railway scheduling systems to achieve higher levels of intelligent control. For example, combined with machine learning algorithms, the BCI system can continuously optimize signal light control strategies, improving overall system efficiency and safety.

In summary, although the application of BCI technology in railway signal light control is still in the exploratory stage, its potential is enormous. With continuous advancements in BCI technology, its application prospects in railway signal light control are broad. In the future, BCI technology is expected to combine with more advanced sensor technologies and artificial intelligence systems to realize fully intelligent railway traffic management. By continuously optimizing technology and system design, BCI technology will provide more precise and efficient solutions for railway transportation.

### 5.1.4 *Brain–Computer Interface and Civil Aviation Signal Light Control*

In China's integrated three-dimensional transportation network, the status of air transportation is continuously rising, making it increasingly important to ensure the efficiency and safety of airport operations. As an emerging intelligent control method, BCI technology can precisely bring an intelligent upgrade to civil aviation signal light control.

The airport signal light system is mainly used to guide aircraft takeoff, landing, and ground taxiing. Due to the complexity of the flight environment and the high safety requirements, traditional signal light control systems face issues such as slow response speed and insufficient flexibility when dealing with emergencies and real-time changes in the flight environment. BCI technology can capture air traffic controllers' EEG signals in real time, enabling faster adjustment of signal light states and enhancing the ability to respond to unexpected situations.

In the civil aviation field, the application of BCI technology mainly targets the complexity and variability of the flight environment. Currently, there are three experimental directions for this technology.

**1. EEG Signal and Environmental Data Fusion:** By combining EEG signals with real-time environmental data (such as weather conditions and runway status), the BCI system can adjust signal light states more intelligently. For example, under adverse weather conditions, the system can dynamically adjust the signal lights based on the air traffic controller's stress level and environmental changes to ensure flight safety.

**2. Multi-dimensional Signal Processing:** In addition to EEG signals, the BCI system can also integrate other physiological signals (such as heart rate and galvanic skin response) to provide a more comprehensive monitoring of the operator's state. This multi-dimensional signal processing can improve the system's response accuracy and promptly detect operator fatigue or distraction.

**3. Real-time Collaborative Decision-making System:** BCI technology works collaboratively with other intelligent airport systems (such as automated ground service equipment and intelligent navigation systems) to form an integrated decision support system. For example, the system can adjust signal lights in real time to coordinate with ground service equipment operations, improving the overall efficiency of airport operations.

Although the practical application of BCI in civil aviation signal light control has not yet been implemented, some related research and experiments have already demonstrated its application potential. Recently, the Insight Data Analytics Center of the Irish Science Foundation released experimental test results on the application of brainwaves in complex environments, emphasizing the unique effectiveness of BCI technology in handling variable environments. This research provides strong theoretical support for the application of BCI technology in the civil aviation field.

Based on the research achievements of BCI technology in the field of transportation, we can see that although the application of this technology in civil aviation signal light control is still at an early stage, its future development potential is enormous, and it can achieve outstanding results in the following aspects.

1. **Large-scale data analysis:** By collecting and analyzing a large amount of EEG signals and flight data, signal processing algorithms can be optimized to enhance the system's intelligence level. This will help improve the accuracy and response speed of signal light control.

2. **Integration of deep learning and artificial intelligence:** Using deep learning and artificial intelligence technologies to conduct deeper analysis and processing of EEG signals, improving the system's predictive capabilities. For example, by learning the operator's operation patterns and behavioral characteristics, potential operational errors can be predicted in advance, and the signal light status can be automatically adjusted.

3. **International standardization:** As the application of BCI technology in the civil aviation field gradually increases, establishing internationally standardized operational protocols and technical standards will help promote and popularize this technology, ensuring its safe application worldwide.

BCI technology provides an innovative and forward-looking solution for civil aviation signal light control. By capturing and interpreting air traffic controllers' EEG signals in real time, the BCI system can achieve intelligent control of airport signal lights, improving the efficiency of flight scheduling and flight safety. Although the application of this technology is still in the exploratory stage, with continuous technological progress and improvement, BCIs will play an increasingly important role in civil aviation management, providing new momentum for the development of modern aviation systems.

## 5.2 The Role of Brain–Computer Interfaces in Traffic Flow Optimization

The rapid development of BCI technology is changing the way traffic flow is optimized. As a technology capable of directly reading and interpreting brain signals, BCIs not only achieve a revolutionary breakthrough in human–machine interaction but also provide new solutions for intelligent traffic management. In traffic flow optimization, BCI technology enables real-time monitoring and analysis of drivers' and traffic managers' EEG signals, allowing the traffic system to dynamically adjust and optimize its operations, thereby significantly improving the efficiency and safety of traffic flow.

The importance of BCI technology in traffic flow optimization is reflected in several aspects. First, BCI technology can monitor the driver's status in real time, including attention, fatigue level, and emotional changes. This information plays a crucial role in adjusting traffic signals and optimizing traffic flow. For example, when the driver feels fatigued, the system can automatically adjust the signal timing to reduce traffic accidents caused by drowsy driving. Second, by combining the driver's EEG signals with data from other traffic sensors, the BCI system can provide more accurate traffic flow predictions and management recommendations, thereby effectively reducing traffic congestion.

In addition, BCI technology also plays a key role in multi-modal transportation coordination. It enables seamless connection between different modes of transportation, improving the overall efficiency and flexibility of the transportation system. In emergency situations, BCI systems can prioritize passage for emergency vehicles, enhancing emergency response capability and efficiency. Personalized transportation services are another important application area of BCI technology, which improves user experience and traffic flow management flexibility by providing customized navigation and parking solutions.

The multi-dimensional applications of BCI technology bring new perspectives and methods to traffic flow optimization. Next, we delve into the specific applications and advantages of BCI in intelligent traffic management systems, multi-modal transportation coordination, and personalized transportation services. These applications not only demonstrate the powerful potential of BCI technology but also point the way for the future development of transportation systems.

### 5.2.1 *Brain–Computer Interface for Traffic Flow Optimization in Intelligent Traffic Management Systems*

In intelligent traffic management systems, BCI technology combines real-time feedback from the human brain with efficient computing capabilities to provide innovative solutions for traffic flow optimization.

Currently, the main application direction of BCI technology in the transportation field is to capture and analyze drivers' EEG signals to obtain real-time information on their attention, fatigue status, and emotional changes. The application effect of BCI has a significant impact on deeply understanding traffic operation conditions. For example, when a driver is in a fatigued state, the intelligent traffic management system can extend red light duration to remind the driver to rest, thereby reducing traffic accidents caused by fatigue driving.

At present, the Massachusetts Institute of Technology (MIT) in the United States has conducted related research on this technology. The main idea of this research is to apply BCI technology combined with vehicle sensor data to monitor the real-time status of drivers. The research direction not only targets improving traffic safety but also focuses on the development of autonomous driving technology. The goal is to integrate autonomous driving technology with BCI technology to enhance vehicle intelligence. That is, when the vehicle detects that the driver is inattentive or fatigued, it can promptly activate the autonomous driving function to ensure the driver's safety while maintaining smooth traffic flow.

From an application perspective, BCI technology can greatly promote intelligent traffic management because drivers' emotions vary significantly under different traffic conditions, and the core of intelligent traffic development is to provide reliable and timely intelligent decision-making for traffic operations. Currently, the development of traffic intelligence mainly targets objective traffic conditions, such as adjusting traffic signals based on traffic volume or providing parking navigation services. However, intelligent traffic services addressing drivers' subjective emotions are rare. The application of BCI technology can enhance development in this area. This technology can deeply analyze drivers' real-time emotions and states (as of 2024, BCI technology cannot yet accurately analyze drivers' complex cognitive goals), which facilitates the deep integration of intelligent traffic management systems with cutting-edge technologies such as big data and AI to provide intelligent services based on drivers' actual needs. For example, when a driver exhibits anxious emotions, the intelligent traffic management system can use BCI technology to determine the driver's needs and offer alternative navigation plans. This approach can improve the application effects of big data, AI, and other technologies, while providing more humanized services in intelligent traffic management.

Regarding traffic flow optimization, the application scope of BCI technology is also very broad. The main current application direction is to more comprehensively utilize traffic resources based on the analysis results of BCI technology. Existing BCI research teams worldwide are already considering how to integrate BCI technology into modern intelligent traffic management systems. Through such technological analysis, traffic management authorities can better understand the actual operation of traffic networks—for example, which roads have drivers generally in an anxious state and which roads have drivers in a more relaxed mental state—and then plan and allocate traffic roads reasonably based on the analysis results to improve the overall utilization rate of each road and alleviate traffic flow pressure.

Overall, BCI technology can enhance the humanization features of modern intelligent traffic management systems, improve road safety and driver experience, and optimize the utilization of traffic resources, thereby alleviating the urban operational pressure caused by traffic flow.

### 5.2.2 *Brain–Computer Interface for Traffic Flow Optimization in Multi-modal Transportation Coordination*

In modern urban transportation systems, the diversity and complexity of transportation modes pose severe challenges to traffic management. The coordination and optimization of various transportation modes such as highways, railways, subways, and aviation are crucial for improving overall traffic flow and reducing congestion. As an advanced interaction method, BCI technology can also facilitate multi-modal transportation coordination, thereby playing an important role in optimizing traffic flow.

Traditional traffic management systems often rely on preset programs and manual operations, making it difficult to respond quickly to emergencies and complex traffic conditions. However, BCI technology can achieve dynamic adjustment of traffic flow through real-time data collection and analysis, enhancing the system's flexibility and response speed. With BCI technology, traffic management centers can monitor and regulate the flow between different transportation modes in real time. For example, when highway traffic volume is excessive, the system can analyze drivers' states through EEG signals and automatically adjust traffic lights to prioritize diverting some vehicles to railway or subway systems, thereby reducing traffic pressure on highways. This cross-system data sharing and collaborative operation not only improve the overall efficiency of the traffic system but also effectively reduce congestion.

In multi-modal transportation systems, BCI technology mainly optimizes traffic flow through the following aspects.

1. **Real-time data sharing and collaborative operation:** The BCI system can integrate EEG signals from drivers and traffic management personnel with data from other traffic sensors to form a comprehensive traffic flow monitoring system. Through real-time data sharing and collaboration, the system can achieve seamless connection between different transportation modes. For example, during peak hours, the system can dynamically adjust traffic lights and the operation of transportation hubs based on changes in EEG signals, optimizing the allocation of traffic resources.

2. **Rapid response in emergencies:** In emergencies, BCI technology can significantly enhance emergency response capabilities. By monitoring the EEG signals of emergency vehicle drivers in real time, the system can quickly identify emergency situations and prioritize adjusting traffic lights to ensure passage for emergency vehicles. For instance, in the event of a traffic accident or emergency rescue mission, the system can swiftly adjust traffic lights along the route or provide lane planning prompts on electronic traffic information screens to ensure emergency vehicles reach the scene as quickly as possible.

3. **Optimization of intelligent transportation hubs:** The application of BCI technology in intelligent transportation hubs can achieve seamless connection between different transportation modes. By installing EEG monitoring devices at transportation hubs such as train stations, subway stations, and airports, the system can monitor passenger flow in real time and dynamically adjust the operation of the hub according to passengers' real-time status and needs. For example, when the system detects a large number of passengers experiencing anxiety, it can optimize crowd guidance to reduce congestion and delays, thereby improving the overall efficiency of traffic flow.

4. **Priority passage for public transportation and special vehicles:** BCI technology can also provide priority passage guarantees for public transportation vehicles and other special vehicles (such as school buses and freight vehicles). By real-time monitoring of the driver's EEG signals, the system can identify these special vehicles and dynamically adjust traffic signals to ensure their rapid passage. This not only improves traffic flow efficiency but also enhances the service quality of public transportation and the travel experience of passengers.

With the continuous advancement of technology, BCI systems have broad application prospects in multi-modal traffic coordination. In the future, BCI systems will further integrate big data and artificial intelligence technologies to achieve more intelligent and automated traffic management.

### 5.2.3 *Traffic Flow Optimization of Personalized Transportation Services by Brain–Computer Interface*

An important goal of the development of intelligent traffic systems is to provide tailored services for each user, thereby improving the overall traffic flow efficiency and user experience. Currently, various cutting-edge technologies such as big data, cloud computing, AI, and blockchain play significant roles in this field, and BCI technology is naturally no exception. Compared to other advanced technologies, BCI technology can more intuitively collect and analyze users' actual needs, thus providing more accurate and flexible human-centered service decisions, which also enhances the intelligence of traffic development. In the following, we analyze in detail how BCI technology optimizes traffic flow by meeting more personalized transportation service demands.[5,6]

First, BCI technology can provide more humanized navigation needs. Based on the application of BCI technology, navigation software can offer the best driving routes according to the driver's real-time status. For example, when the system detects that the driver's EEG signals indicate fatigue, it can automatically adjust the navigation route to avoid congested areas and recommend the nearest rest stops. This real-time adjustment not only reduces driver fatigue but also optimizes overall traffic flow, preventing traffic jams caused by sudden situations.

---

[5]Cloud computing is a form of distributed computing that involves breaking down large-scale data processing programs into numerous smaller tasks via the internet-based "cloud." These tasks are then processed and analyzed by systems composed of multiple servers, with the results subsequently returned to the user.

[6]A blockchain is a decentralized, distributed ledger that employs a chain-of-blocks structure for data storage, ensuring immutability, security, and trustworthiness. It integrates technologies such as distributed storage, peer-to-peer transmission, consensus mechanisms, and cryptography. By recording transactions and information within sequentially linked and continuously expanding data blocks, it guarantees data security and transparency.

In addition, BCI technology can combine real-time traffic data to provide dynamic route suggestions for drivers. For instance, during peak traffic hours, the system can offer alternative routes based on traffic conditions and the driver's status, reducing waiting times and congestion. This dynamic navigation approach can significantly improve traffic flow efficiency and enhance the driver's travel experience.

Moreover, parking issues in city centers have always been a major challenge in traffic management. BCI technology, by monitoring the driver's status in real time, can optimize the allocation and use of parking spaces. For example, the system can pre-book parking spots based on the driver's EEG signals as they approach their destination and guide them along the optimal parking route. Such intelligent parking management not only improves parking efficiency but also reduces traffic congestion caused by searching for parking spaces, thereby improving traffic flow in city centers.

Finally, BCI technology can also be applied to public transportation services to enhance passenger experience and optimize traffic flow. The main method is through real-time monitoring of passengers' EEG signals, allowing urban bus dispatch systems to dynamically adjust public services such as bus and subway departure frequencies and route arrangements. Passengers' emotions and needs are directly reflected in EEG signals, and the system uses this information to provide personalized service recommendations and information prompts, thereby improving passenger satisfaction.

Of course, although BCI technology has broad application prospects in personalized transportation services, it also faces some challenges. Data privacy and security are primary concerns; how to protect users' EEG data from misuse is a problem that must be addressed during technological development. Additionally, the reliability and stability of the technology need further improvement to ensure the system can operate stably in complex and variable traffic environments.

It is believed that with the continuous advancement of artificial intelligence and big data technologies, BCI systems will achieve higher intelligence and automation in personalized transportation services and deeply integrate into people's daily lives, providing more travel solutions tailored to actual needs. Therefore, it is imaginable that the multi-dimensional application of BCI technology in the future will make significant contributions to personalized services and traffic flow optimization in global traffic systems.

## 5.3 Application of Brain–Computer Interface in Traffic Accident Prevention

Traffic safety is the top priority in traffic development, and leveraging technological power to enhance traffic safety assurance has always been a key strategy for our high-quality traffic development. In the integration process of BCI technology and traffic development, this technology is providing new approaches and means to improve road traffic safety and reduce the occurrence of accidents.

The core advantage of BCI technology lies in its ability to directly read and interpret human brain signals, providing feedback that is more precise and immediate than traditional sensors and monitoring devices. Whether in driving state monitoring, integration of accident prevention technologies, or in education, training, and accident analysis, BCI demonstrates tremendous potential and application value. The introduction of this technology not only helps enhance drivers' safety awareness and driving skills but also provides traffic management authorities with more scientific data support and decision-making basis.

With continuous technological advancement, the application prospects of BCI in traffic accident prevention are broad. In the future, as artificial intelligence, big data, and the Internet of Things further integrate, BCI systems will become more intelligent and automated, providing strong technical support for building safer and more efficient traffic systems.

### 5.3.1 *Brain–Computer Interface Monitoring of Driving State*

BCI technology offers a novel method for monitoring driving states, overcoming the limitations of traditional monitoring approaches. By real-time monitoring of drivers' EEG signals, BCI technology can accurately identify drivers' attention levels, fatigue states, and emotional changes, enabling in-depth monitoring that is highly effective in preventing traffic accidents.

Concentration is one of the fundamental prerequisites for safe driving. Traditional driving state monitoring methods, such as cameras and sensors, can capture drivers' facial expressions and head movements but often fail to deeply understand the drivers' internal psychological states. BCI technology, by real-time monitoring of EEG signals, can precisely identify drivers' attention levels. For example, when a driver's attention begins to decline, specific changes appear in the EEG signals, and the BCI system can promptly issue alerts to remind the driver to regain focus, thereby preventing traffic accidents caused by distraction.

Fatigue driving is one of the main causes of traffic accidents. Traditional fatigue monitoring methods, such as camera monitoring of eye closure or head position, are somewhat effective but still suffer from latency and insufficient accuracy. BCI technology can identify drivers' fatigue states in advance by monitoring slow-wave activity and other fatigue-related signals in EEG. Studies show that when drivers feel fatigued, slow-wave activity in EEG increases, and changes in $\alpha$ waves and $\theta$ waves also become apparent. The BCI system can issue rest recommendations based on these signals in advance, thereby preventing traffic accidents caused by fatigue driving.[7,8]

The emotional state of drivers also directly affects driving safety. Negative emotions such as anger, anxiety, and stress may lead to dangerous driving behaviors. BCI technology can monitor drivers' emotional changes in real time by analyzing specific frequency band activities in EEG signals to identify their emotional states. For example, when a driver experiences significant emotional fluctuations, $\beta$ wave activity in EEG significantly increases. The BCI system can use these signals to remind drivers to calm their emotions or suggest temporarily stopping to rest, thereby avoiding traffic accidents caused by emotional loss of control.

---

[7]The $\alpha$ wave is one of the four fundamental brainwave patterns. It is commonly associated with the subconscious state, which occurs when the human brain exhibits $\alpha$ wave activity. Serving as a bridge between the conscious and subconscious mind, the $\alpha$ wave provides the only effective pathway to access the subconscious. It facilitates inspiration, accelerates information absorption, enhances memory, and is considered the optimal brainwave state for learning and thinking. When the brain is dominated by $\alpha$ waves, conscious mental activity is significantly inhibited, rendering logical thinking and reasoning difficult. During this state, the brain processes and transmits information primarily through intuition, inspiration, and imagination.

[8]Theta waves refer to a component of EEG with a duration of 1/8–1/4 seconds. The brainwave rhythm with a frequency of 4–8 Hz is termed theta rhythm. In normal adult EEGs, a small amount of theta waves is typically present, with a frequency of 4–7 Hz and an amplitude of 20–40 microvolts. These waves are sporadically distributed, primarily observed in the frontal and temporal regions, where their index is less than 20%, while in other areas, the index is below 10%. Theta waves with higher voltage in adults are generally considered pathological.

Currently, the application of BCI technology in driving state monitoring has been validated in some trials and studies. For example, as mentioned above, DiYi Technology's development of the "Portable BCI Driving Safety Intelligent Prevention and Control System" in the Shanxi Transformation and Comprehensive Reform Demonstration Zone has achieved remarkable results, demonstrating the unique advantages of BCI technology in reducing traffic accidents.

However, although the application value of BCI technology has been confirmed, some challenges remain in practical use, such as noise interference in EEG signals, comfort of device wearing, and system real-time performance. To overcome these challenges, researchers are continuously optimizing BCI system algorithms to improve the accuracy and speed of signal processing. At the same time, new EEG signal acquisition devices are being developed to enhance comfort and portability. For example, the emergence of lightweight EEG caps and non-invasive EEG sensors has made the application of BCI technology in actual driving more feasible.

It is believed that in the near future, the application of BCI technology in driving state monitoring will provide a revolutionary tool for traffic accident prevention. By real-time monitoring and analyzing the driver's EEG signals, accurately identifying attention, fatigue, and emotional states, timely warnings and corresponding measures can be taken, thereby significantly improving driving safety and making an important contribution to building a safer traffic system.

### 5.3.2 *Integration of Brain–Computer Interface in Accident Prevention Technologies*

The application of BCI technology in traffic accident prevention is not limited to monitoring a single driving state; it can also build a comprehensive intelligent traffic safety system through integration with other prevention technologies. By combining multiple advanced technologies, BCI can provide more comprehensive and precise accident prevention solutions, thereby significantly enhancing road traffic safety. The specific applications can be reflected in the following aspects.

**1. Integration of BCI with Warning Systems:** In accident prevention, the integration of BCI technology with warning systems is one of its key application directions. BCI technology can monitor the driver's EEG signals in real time to promptly detect potential risk factors and quickly issue alerts through the warning system. More importantly, this warning system can be further integrated with automatic intervention systems to directly intervene in the driving process when necessary. For example, if the driver fails to respond timely due to inattention, the system can automatically activate braking or steering operations, or even initiate autonomous driving functions to avoid collision accidents.

**2. Integration of BCI with Sensing Technologies:** The fusion of BCI technology with other sensing technologies can achieve comprehensive accident prevention. By combining EEG signals with data from vehicle sensors such as radar, LiDAR, and cameras, the vehicle's intelligent system can obtain a more comprehensive environmental perception capability. This multi-sensor fusion technology can provide precise environmental information and driver status, thereby better predicting and preventing potential traffic accidents. For example, when radar detects an obstacle ahead and the BCI system simultaneously identifies driver inattention, the system can issue an early warning and take braking measures to avoid a collision.[9,10]

---

[9]Sensing technology, alongside computing and communication, is recognized as one of the three pillars of information technology. From the perspective of the Internet of Things (IoT), sensing technology is a crucial indicator of a nation's informatization level. Events such as the "2nd Hangzhou IoT and Sensing Technology Application Summit Forum" are instrumental in promoting the rapid industrial development of sensors in China. As a modern interdisciplinary science and engineering discipline, sensing technology focuses on acquiring information from natural sources, followed by processing (transforming) and recognizing it. It encompasses a range of activities including the planning, design, development, manufacturing/construction, testing, application, and evaluation/improvement of sensors (also known as transducers), as well as related information processing and recognition systems.

[10]Lidar (Laser Radar) is a radar system that emits laser beams to detect characteristics of targets such as position and velocity. Its operational principle involves transmitting a detection signal (laser beam) toward a target, then comparing the received reflected signal (target echo) with the transmitted signal. After appropriate processing, relevant information about the target—such as distance, azimuth, altitude, speed, attitude, and even shape—can be obtained. This enables the detection, tracking, and identification of targets such as aircraft and missiles.

**3. Integration of BCI with Intelligent Traffic Infrastructure:** The integration of BCI technology into intelligent traffic infrastructure can enable more intelligent traffic management. By combining the driver's EEG signals with traffic signal control systems, smart roads, and Vehicle-to-Everything (V2X) technology, the traffic management system can dynamically adjust signal timing, lane usage, and traffic flow to prevent accidents. For example, when the system detects multiple drivers experiencing fatigue on a certain road segment, it can automatically adjust the signal timing in that area to slow down traffic flow and reduce accident risks.

**4. Optimization of Human–Machine Interaction:** BCI technology can also improve accident prevention effectiveness by optimizing human–machine interaction. Although traditional Advanced Driver Assistance Systems (ADAS) provide various assistance functions, their interfaces and alert methods often fail to capture the driver's attention at critical moments. BCI technology, by directly reading the driver's EEG signals, can achieve more intuitive and efficient human–machine interaction. For example, the system can adjust the presentation style and timing of alerts based on the driver's real-time state, making them more aligned with the driver's cognitive habits, thereby enhancing the effectiveness and timeliness of warnings.[11]

Currently, the integrated application of BCI technology in accident prevention has been experimentally studied and researched worldwide. China also has numerous research projects, such as the "Embedded RSVP System Research Based on BCI" conducted by Beijing Jiaotong University and the "Automotive Cabin Sound and Vibration Experience Research Based on BCI" by Southwest Jiaotong University, all of which are promoting the application of this technology in the transportation field.

Of course, although the current research directions are clear, there are still some challenges in practical applications. For example, most existing research results need further improvement in reliability and real-time performance to ensure stable operation in various complex traffic

---

[11] Driving assistance systems include the Lane Keeping Assist System, Automatic Parking Assist System, Brake Assist System, Reverse Assist System, and Driving Assist System.

environments. Additionally, the comfort and user acceptance of related research devices also require continuous enhancement to better integrate into daily driving.

However, the overall current development trend has demonstrated that BCI technology, through integration with various accident prevention technologies, provides new ideas and methods for building comprehensive intelligent traffic safety systems. With continuous technological advancement and optimization, the application prospects of BCI technology in accident prevention will become increasingly broad, bringing greater security guarantees to the safe development of intelligent transportation systems.

### 5.3.3 *Application of Brain–Computer Interface in Education and Training*

Traffic driving safety is closely related to individual driving habits, and personal driving habits are directly connected to the learning of driving skills. Therefore, improving the effectiveness of driving skills education and training can also enhance traffic safety, and BCI technology can be precisely applied in this field.

The application of BCI technology in education and training can provide support for improving drivers' skills and safety awareness. By real-time monitoring and analyzing trainees' EEG signals, BCI technology can not only enhance the effectiveness and efficiency of training but also provide a scientific basis for personalized teaching and training program development.

Driving simulators have now become widely popular in driver training. However, traditional driving simulators mainly rely on visual and operational feedback, making it difficult to fully understand trainees' psychological states and reaction abilities. BCI technology, by real-time monitoring of trainees' EEG signals, can gain deeper insights into their attention, stress, and emotional states, thereby providing more comprehensive feedback and guidance.[12]

---

[12]A car driving simulator cockpit is composed of a driver's seat, a visual simulation computer, a display screen (19-inch monitor), operation sensors, a data acquisition card, headphones, and a microphone. The cockpit incorporates operational components identical to those of a real vehicle, including the five major control mechanisms: steering wheel,

The application of BCI technology in driving simulation training can monitor trainees' EEG signals in real time and dynamically adjust training difficulty and content based on their attention levels and emotional changes. For example, when the system detects that a trainee's attention is distracted, it can increase the training difficulty to improve their concentration ability; when excessive stress is detected, it can appropriately reduce training intensity to avoid negatively impacting learning outcomes due to over-tension. This personalized training approach helps improve trainees' driving skills and their ability to handle complex traffic situations. By analyzing EEG signals, BCI technology can assist instructors in better understanding trainees' learning status and progress, providing targeted guidance and suggestions to enhance the effectiveness of driver education and training.

In addition to strengthening driving skills, BCI technology is also widely applied in safety awareness training. By simulating various driving scenarios and emergencies, BCI systems can help trainees experience and learn how to respond to risks and challenges in real driving within a safe environment. BCI technology can set up various sudden events and hazardous scenarios during simulation training, evaluate trainees' risk perception and response abilities through real-time monitoring of EEG signals, and provide timely guidance to help drivers enhance risk awareness and coping skills, fostering good driving habits.

Moreover, the application of BCI technology in simulated driving allows trainees to repeatedly practice various preventive measures and emergency operations in different hazardous environments, strengthening their ability to respond to sudden incidents. This data-driven rehearsal method helps improve trainees' emergency handling skills and safety awareness in real driving situations.

In summary, through BCI technology, any vehicle driving instructor and training institution can effectively enhance the training and education

clutch, foot brake, accelerator, and handbrake. It features a realistic gearbox with reverse gear, first gear, second gear, third gear, fourth gear, fifth gear, and neutral gear (automatic transmission models include only drive gear, reverse gear, and park gear). Authentic vehicle control switches are also provided, such as left turn signal, right turn signal, hazard lights, horn, ignition switch, main power switch, seat belt, door control, windshield wipers, high beam, low beam, and high-low beam alternation. The simulator cockpit supports both networked training and standalone training sessions.

of drivers, thereby cultivating more excellent drivers and improving traffic safety from the basic operational level.

### 5.3.4 *Application of Brain–Computer Interface in Accident Analysis*

BCI technology plays an important role not only in preventing traffic accidents but also in accident analysis, thereby providing a scientific basis for improving traffic safety measures.

In traffic accident analysis, understanding the driver's state before the accident occurs is crucial. Traditional accident analysis methods usually rely on on-site investigation and post-accident inquiry, making it difficult to fully grasp the driver's psychological state and behavioral process. BCI technology can record the driver's attention, fatigue level, and emotional changes before the accident by real-time monitoring of the driver's EEG signals. This data is of great value for understanding the causes of accidents. For example, by analyzing the EEG waves from several minutes before the accident, researchers can determine whether the driver was sluggish due to fatigue or failed to take timely action due to distracted attention.

At the moment of the accident, the driver's reaction speed and decision-making ability directly affect the outcome. BCI technology can record the driver's EEG activity during the accident, analyzing their reaction process and decision-making path. By analyzing this data, it is possible to evaluate the driver's reaction speed and decision quality in emergency situations. This in-depth analysis helps understand the driver's specific performance during the accident, thereby providing a basis for improving driver training and safety education.

By collecting and analyzing a large amount of accident data, BCI technology can provide data support for the improvement of traffic safety measures. For example, through the analysis of EEG signals from multiple similar accidents, common accident causes and typical driver response patterns can be identified. This information can be used to improve traffic safety regulations and driver training programs. Additionally, if it is found that most accidents are caused by drivers' lack of attention in specific environments, stricter traffic regulations can be formulated for such situations, or relevant scenario-based simulation training can be added to driver education.

Therefore, BCI technology can analyze the influencing factors of accidents from the driver's subjective perspective, providing new practical approaches for in-depth research on traffic accidents. It helps improve traffic safety measures and driver training programs, and also provides a scientific basis for enhancing the overall safety of the traffic system.

# Chapter 6

# Brain–Computer Interface and the Operation of Vehicles, Ships, and Aircraft

## 6.1 Brain–Computer Interface and Autonomous Driving Technology

As mentioned in the previous chapter, the integration of brain–computer interface (BCI) technology with autonomous driving technology is not only a trend in the development of modern intelligent transportation but also an important opportunity for upgrading smart transportation. This integration can enhance driving safety and system decision-making capabilities, as well as open up more intelligent and personalized driving experiences.

Currently, although autonomous driving technology in China is developing rapidly, it has not yet reached the fully autonomous Level 5 as of 2024. Among the autonomous vehicles in operation, the Apollo Moon, a deeply customized driverless vehicle jointly developed by Baidu and BAIC Arcfox, is the highest-level autonomous vehicle model available. This model has also pioneered a new autonomous driving brand in the transportation sector called "Luobo Kuaipao" (Radish Express). Figure 6.1 is an image of Luobo Kuaipao. However, traditional autonomous driving technology remains primarily assistive, operating only within designated areas and relying on technologies such as big data and cloud computing. It has not yet achieved full autonomy and cannot meet the customization needs of drivers (passengers). BCI technology, on the

**Figure 6.1.** Baidu's autonomous ride-hailing service: Radish Rush (Apollo Go).

other hand, can fill this critical technological gap and enhance the safety and flexibility of autonomous driving.[1,2]

[1]Apollo Moon is the fifth-generation autonomous vehicle released by Baidu Apollo. It features a new suite of autonomous driving hardware, with an overall cost of 480,000 RMB, which falls within the operational cost range of ride-hailing services, and it possesses full self-driving capabilities. Leveraging Baidu Apollo's globally leading autonomous driving technology, Apollo Moon adopts the "ANP-Robotaxi" architecture. This design not only reduces the weight of the autonomous vehicle hardware but also enables data sharing and synergy with intelligent driving vehicles, creating a powerful closed-loop data system. By adding one customized lidar and corresponding redundancy for autonomous driving on top of the Navigation Assisted Driving (ANP) model, full self-driving capability is achieved.

[2]Apollo Go is an autonomous vehicle mobility service platform that has initiated public ride-hailing testing and operations in 11 cities across China, achieving full coverage in all tier-one megacities. Furthermore, the service has commenced fully driverless autonomous ride-hailing operations and testing in Beijing, Wuhan, Chongqing, Shenzhen, and Shanghai. As of April 19, 2024, Apollo Go has completed over 6 million cumulative rides on public roads, solidifying its position as the world's largest provider of autonomous mobility services.

From the trend perspective, with the rapid development of artificial intelligence, big data, and Internet of Things technologies, the integration of BCI and autonomous driving technologies will become increasingly close. In the future, this integration can not only be applied to private vehicles but will also play an important role in public transportation, logistics, and other fields. BCI technology, by enhancing the intelligence level of autonomous driving systems and the human–machine interaction experience, will promote the popularization and application of autonomous driving technology, making a significant contribution to building a safer and more efficient intelligent transportation system.[3]

Moreover, the application of BCI technology in autonomous driving also holds important social significance. It not only improves road traffic safety and reduces traffic accidents but also enhances user satisfaction and comfort by providing a personalized driving experience. With continuous technological advancement, the combination of BCI and autonomous driving technologies will gradually transition from laboratory research to practical application, painting a brighter blueprint for the future development of intelligent transportation.

### 6.1.1 *The Role and Advantages of Brain–Computer Interface in Autonomous Driving*

The BCI provides real-time feedback on the driver's status and control capabilities by directly reading and interpreting the driver's EEG signals, enhancing the decision-making ability and safety of autonomous driving systems. This can bring a qualitative change to the development of autonomous driving technology. From this perspective alone, BCI technology holds unique advantages and value in autonomous driving.

As mentioned earlier, BCI technology can monitor the driver's EEG signals in real time to assess their attention level, fatigue state, and emotional changes. This real-time monitoring is especially important during the autonomous driving process, particularly at the current stage of autonomous driving development, where this technology has broader

[3] The Internet of Things (IoT) originated in the field of media and represents the third revolution in the information technology industry. It refers to the interconnection of any object with a network through information sensing devices and established protocols, enabling objects to exchange information and communicate via information transmission media. This facilitates intelligent identification, localization, tracking, monitoring, and related functions.

application potential. Since current autonomous driving technology mainly manifests as driver assistance, this technology reduces the frequency of driver operations, providing certain convenience but also causing a decline in driver attention. Therefore, if the vehicle's autonomous driving system can use BCI technology to monitor in real time, it can match corresponding driver assistance services based on the driver's actual condition.

Take the emergency evasive autonomous driving function equipped in many current vehicles as an example. Most vehicles' emergency evasive functions are actually a type of brake assist system. It mainly provides additional braking force at critical moments by analyzing the driver's braking behavior in detail, significantly shortening the braking distance to achieve emergency avoidance. Although this function respects the driver's judgment during normal driving, its intervention timing usually only activates when the vehicle's data indicates a critical situation, meaning the driver and passengers are already facing danger. However, after integrating BCI technology with autonomous driving technology, a different effect can be produced. It can add the driver's status data under the above premise, comprehensively analyze the probability of dangerous situations, greatly predict potential hazards, and then proactively avoid various dangers through voice or text reminders or by intervening in driving.

Currently, with the deepening research and application of BCI technology, its integration with autonomous driving technology is continuously advancing. This technology will promote the popularization and application of autonomous driving technology, building a safer and more efficient intelligent transportation system.

### 6.1.2 *Integration of Brain–Computer Interface and Vehicle Control Systems*

The integration of BCI technology with autonomous driving technology is mainly reflected in its integration with vehicle control systems, which is also key to achieving intelligent and personalized vehicle upgrades. This integration not only improves the response speed and driver assistance capabilities of autonomous driving systems but also provides drivers with more flexible and personalized control methods.

From a theoretical perspective, the neural feedback mechanism of BCI technology can optimize the vehicle's control algorithms by analyzing the driver's EEG signals in real time. By monitoring the driver's brainwaves, the vehicle's autonomous driving system can identify the driver's driving habits and preferences, thereby adjusting the vehicle's speed, steering, and braking strategies. This personalized control approach can enhance the comfort of both the driver and passengers, as well as provide faster responses in emergency situations.

Additionally, BCI technology enables the vehicle's autonomous driving capabilities to better align with the driver's driving habits, thereby enhancing the driving and riding experience. Since BCI technology can analyze driving data over a long period, it identifies the driver's specific behavior patterns and optimizes control algorithms accordingly, thus improving driving decisions. In other words, BCI technology allows the driver to feel more like they are driving the vehicle themselves during autonomous driving, which in turn increases the driver's satisfaction with autonomous driving.

Finally, BCI technology can also be combined with augmented reality (AR) technology to provide drivers with more personalized information prompts. By monitoring the driver's EEG signals in real time, the system can display the most relevant information on the AR display, improving the convenience and safety of the driving experience. For example, when the driver is distracted, the system can highlight obstacles and road conditions ahead on the AR display to help the driver react promptly.[4]

From a hardware perspective, the integration of BCI technology with other sensor technologies can significantly enhance the environmental perception capabilities of autonomous driving systems. By combining the driver's EEG signals with data from LiDAR, cameras, and ultrasonic sensors, the system can obtain more comprehensive and accurate environmental information. This multi-sensor data fusion technology improves the perception ability of autonomous driving systems and provides more

---

[4]Augmented Reality (AR) technology integrates computer-generated virtual environments with the user's real-world surroundings by leveraging optoelectronic display technology, interactive technology, various sensor technologies, and computer graphics and multimedia systems. This integration enables users to perceive the virtual environment as a seamless part of their physical reality through sensory effects. AR is characterized by the fusion of virtual and real elements, real-time interaction, and three-dimensional registration.

precise decision support in complex road conditions. For instance, when radar detects an increase in traffic flow ahead and the BCI system simultaneously identifies that the driver's attention is insufficient, the system can issue an early warning and take corresponding measures to prevent potential dangers.

The integration of BCI technology with vehicle control systems is a major focus in current autonomous driving research and development. Through control algorithm optimization and multi-sensor data fusion, BCI technology can significantly enhance the intelligence level and driving experience of autonomous driving systems. This technology has broad application prospects and development space in the field of autonomous driving and is a key technology to accelerate the widespread adoption and implementation of autonomous driving.

### 6.1.3 *Main Directions of Integration between Brain–Computer Interface and Autonomous Driving*

Today, the application of BCI technology in autonomous driving shows broad prospects. Leveraging its unique advantages and characteristics, BCI technology is providing more possibilities and innovative directions for autonomous driving systems. Analyzing the current state of integration between this technology and autonomous driving reveals that the fusion mainly manifests in the following directions.

#### 1. Human–Machine Co-Driving Mode

The human–machine co-driving mode is one of the important directions in the integration of BCI and autonomous driving technologies. In the current stage where autonomous driving technology mainly plays an auxiliary role, drivers still need to maintain a certain level of attention to take over control when necessary. BCI technology can achieve seamless switching between autonomous and manual driving by monitoring the driver's EEG signals in real time. When the system detects that the driver is focused, it can automatically switch to manual driving mode; otherwise, it switches back to autonomous driving mode. This seamless switching mechanism not only improves driving safety but also enhances the driver's sense of control over the vehicle.

Additionally, during autonomous driving, the BCI system can provide assistance functions to the driver. When the driver encounters complex road conditions or emergencies in autonomous mode, the BCI system can monitor their reactions through EEG signals and provide timely assistance. This human–machine collaboration effectively reduces the driver's burden and improves both the driving experience and safety.

**2. Combination of Fully Autonomous and Semi-Autonomous Driving**
In semi-autonomous driving mode, the application of BCI technology can significantly improve driving safety. By monitoring the driver's EEG signals, the system can understand their attention and fatigue status in real time and provide assisted control when necessary. In fully autonomous driving mode, BCI technology is mainly used to monitor and respond to emergencies. Although fully autonomous vehicles have a high degree of autonomy, driver involvement is still required in certain complex or urgent situations. The BCI system can quickly awaken the driver and provide necessary assisted operations in emergencies by real-time monitoring of the driver's EEG signals.

**3. Brain–Computer Interface and Intelligent Transportation Infrastructure Collaboration**
Autonomous driving technology does not only require the installation of corresponding hardware on vehicles but also necessitates technical upgrades to transportation infrastructure. The integration of BCI technology with intelligent traffic signals is key to the comprehensive implementation of autonomous driving technology.

Transportation infrastructure can collaborate with intelligent traffic signal systems by monitoring drivers' EEG signals, dynamically adjusting the timing and frequency of traffic lights based on real-time data to optimize traffic flow and reduce congestion.

BCI technology can also enhance the intelligence level of autonomous driving systems through integration with Vehicle-to-Everything (V2X) and Internet of Things (IoT) technologies. By real-time monitoring and data sharing, BCI systems can exchange information with other vehicles and infrastructure to achieve global optimization and coordinated operation. For example, in large logistics fleets, BCI systems can monitor each driver's status in real time and dynamically adjust vehicle routes and speeds based on overall traffic conditions and driver states, improving logistics efficiency and safety.

As of August 2024, the integration of BCI technology and autonomous driving technology has been deeply researched worldwide, with multiple projects entering the implementation phase.

In March 2024, Tesla CEO Elon Musk announced that Neuralink, his BCI company, successfully completed the world's first human brain implant surgery, with the volunteer currently recovering well. The company's first product is named "Telepathy," which allows control of devices such as phones and computers through thought after brain implantation. The company stated that full integration of "Telepathy" with autonomous driving is still some distance away, but if "Telepathy" can comprehensively and accurately capture EEG signals related to driving actions and intentions, autonomous driving technology will undergo a significant transformation. With the support of BCI technology, in the future, driving by thought alone will become possible.

As early as 2022, Ford announced a BCI patent: chassis input intention prediction through the fusion of BCI and driver monitoring sensors. The patent describes "a method of controlling a vehicle using a BCI device." BCI technology helps vehicle systems predict driver behavior by monitoring brainwave activity, thereby assisting vehicle driving. However, currently, this technology can only enhance driving behavior and cannot fully replace driver operation.

All signs indicate that the integration of BCI technology and autonomous driving technology is inevitable. Future autonomous driving will not only involve vehicle automation but also provide diversified and customized driving services based on driver habits, emotions, and even individual thoughts.

## 6.2 Brain–Computer Interface and Driver Monitoring and Assistance Systems

In intelligent traffic safety management, driver status monitoring and assistance systems play a crucial role. With the continuous development of autonomous driving technology, drivers remain a key component of the traffic system. The introduction of BCI technology is precisely the key to resolving conflicts between autonomous driving technology and poor driving habits.

BCI technology can accurately identify drivers' attention levels, fatigue states, and emotional changes by real-time monitoring and analysis of their EEG signals. This real-time monitoring not only provides immediate feedback to alert drivers to stay vigilant but also enables intervention when necessary to prevent potential dangers. Compared to traditional monitoring methods, BCI technology offers higher precision and faster response times, thereby more effectively enhancing driving safety.

In advanced driver-assistance systems, the application of BCI technology further enhances the system's intelligence level. By integrating with lane-keeping, blind-spot monitoring, and automatic emergency braking systems, BCI technology can provide more personalized and precise driving assistance functions.

### 6.2.1 *Driver Monitoring by Brain–Computer Interface*

The driver is the primary subject of BCI technology applications across various transportation modes. The main focus is not on vehicles, ships, or aircraft themselves but on the driver, whose driving decisions are all based on attention levels, fatigue states, and emotional changes. Therefore, the effectiveness of BCI technology in monitoring the driver ultimately determines its overall application outcome.

From the application perspective of BCI technology, the data currently collected in the transportation field mainly targets three aspects: attention level, fatigue state, and emotion. We can develop application strategies for BCI technology based on these three aspects.

**1. Attention Monitoring**

The driver's level of attention concentration is a fundamental guarantee of driving safety. BCI technology can accurately assess the driver's attention state by real-time monitoring of EEG signals. Specifically, the BCI system collects EEG signals through electrodes and analyzes brainwave patterns related to attention, such as changes in alpha ($\alpha$) and beta ($\beta$) waves. When the driver is focused, beta wave activity increases; when attention is distracted, alpha wave activity rises.

When the system detects a decline in the driver's attention, it automatically issues an alert to remind the driver to refocus or takes other intervention measures. For example, when a ship is equipped with a BCI-integrated driver assistance system, if the water surface environment changes or the driver cannot promptly perceive the actual water conditions, BCI technology can provide relevant alerts based on the driver's attention status, reducing traffic accidents caused by subjective negligence.

**2. Fatigue Monitoring**

Fatigue driving is one of the major causes of traffic accidents. Traditional fatigue monitoring methods, such as camera-based monitoring of eye closure or head position, are somewhat effective but still suffer from latency and insufficient accuracy. BCI technology can detect the driver's fatigue state in advance by monitoring slow-wave activity and other fatigue-related signals in EEG. For example, during fatigue, theta ($\theta$) and delta ($\delta$) wave activities increase, while alpha ($\alpha$) wave activity decreases. By analyzing these signal changes, the system can identify the driver's fatigue state.

**3. Emotion Monitoring**

The emotional state of the driver has a direct impact on driving behavior. Negative emotions such as anger, anxiety, and stress may lead to dangerous driving behaviors. BCI technology can identify emotional changes by real-time monitoring of the driver's EEG signals and provide corresponding regulation measures. For example, when emotional fluctuations are significant, the activity of $\beta$ waves in the EEG increases while the activity of $\alpha$ waves decreases. By analyzing these signals, the system can recognize the driver's emotional state.

When the system detects that the driver's EEG shows significant emotional fluctuations, it can automatically play soothing music or suggest deep breathing exercises to help calm the driver. This emotion monitoring and regulation function not only helps reduce traffic accidents caused by emotional loss of control but also improves the overall driving experience and mental health of the driver.

Currently, the practical application of BCI technology in driver monitoring has achieved certain results. For example, in Japan, Nissan Motor Corporation is researching a technology called "Brain-to-Vehicle" (B2V), which predicts the driver's driving intentions and controls the vehicle accordingly by real-time monitoring of the driver's EEG. This technology can adjust the vehicle before the driver performs actual operations and can also adjust the seat, temperature, and other settings based on the driver's EEG signals to enhance driving comfort and safety. Experimental results show that B2V technology performs excellently in reducing reaction time and improving driving accuracy, demonstrating the great potential of BCI technology in real driving environments.

Overall, the application of BCI technology in driver monitoring provides a new means to improve driving safety and user experience. By real-time monitoring of the driver's attention, fatigue, and emotional state, BCI technology can offer effective warnings and interventions, significantly reducing the occurrence of traffic accidents.

### 6.2.2 *Brain–Computer Interface and Driver Assistance Systems*

In intelligent transportation, driver assistance systems have become important tools for enhancing safety and operational efficiency. Traditional driver assistance systems rely on sensors and preset programs, making it difficult to understand the operator's real-time status and needs. The introduction of BCI technology can help the assisted driving systems of vehicles deeply understand the driver's real-time needs and then combine big data, cloud computing, AI, and other technologies to make more precise assistance decisions, greatly promoting the capability improvement of assistance systems.

Currently, the integration and development of BCI technology with driver assistance systems have been carried out on multiple levels. A review of existing research shows that the application of BCI technology in driver assistance systems has achieved remarkable results in various fields.

**1. Vehicle Driver Assistance Systems**

In vehicle operation, the lane-keeping assistance system is a key function to ensure the vehicle stays in the correct lane. As mentioned multiple times before, BCI technology can monitor the driver's attention and fatigue status in real time and provide timely lane departure warnings.

In addition, the blind spot areas of vehicles are dead zones in the driver's field of vision, which can easily lead to traffic accidents. BCI technology, by combining EEG signals and vehicle sensor data, can monitor the vehicle's blind spot areas in real time. When the system detects potential danger in the blind spot area that the driver has not noted, it can issue an alert to remind the driver to pay attention to the blind spot. At the same time, the system can also provide assisted operations when necessary, such as automatic braking or steering adjustments, to avoid collision accidents.

There is also an automatic emergency braking system. The application of BCI technology in automatic emergency braking systems can effectively prevent collision accidents. By integrating EEG signal monitoring with other sensor data, the system can identify potential collision hazards ahead. When the driver fails to respond in time, the BCI system can automatically perform emergency braking operations to avoid or reduce the severity of collision accidents.

**2. Ship Operation Assistance System**

In ship operation, besides providing assisted driving services based on the operator's status, the main function is to give relevant reminders or driving interventions to the personnel in the cockpit according to the ship's navigation environment. For example, in complex environments such as nighttime or storms, the BCI system issues early warnings to the driver based on hazard factors detected by the ship navigation detection system. Or in busy port areas, it helps the driver promptly detect and avoid other vessels, reducing collision risks. From the application logic perspective, the use of brain–computer interfaces (BCIs) in ship operation systems is similar to that in the vehicle field, but it can provide faster and more comprehensive assistance in coping with complex environments.

**3. Aircraft Flight Assistance System**

During flight, aircraft need to continuously monitor the surrounding airspace to avoid collisions with other aircraft or obstacles. BCI technology, by combining EEG signals and aircraft sensor data, can monitor the surrounding airspace in real time. When the system detects potential collision hazards, it can issue alerts to remind the pilot and, if necessary, perform automatic collision avoidance operations to ensure flight safety.

Overall, the application of BCI technology in driving assistance systems provides a new approach to improving operational safety and efficiency. With continuous technological advancement, BCI technology will play an increasingly important role in driving assistance systems, offering more convenience for their development.

## 6.3 Brain–Computer Interfaces and the Comfort and Convenience of Vehicles, Ships, and Aircraft

As people's demands for travel experience continue to rise, there is enormous potential and development direction for improving the comfort and convenience of transportation tools. For example, in terms of comfort, users expect more personalized and intelligent environmental adjustments; in terms of convenience, users hope to simplify operation processes and achieve more efficient and easier control and information access. BCI technology provides unprecedented solutions for meeting these needs by real-time monitoring and interpreting users' EEG signals.

In these improvement directions, BCI technology, through its unique real-time monitoring and feedback mechanisms, further promotes the advancement of comfort and convenience in transportation tools. This technology will play an increasingly important role in the design and use of transportation tools, providing users with unprecedented intelligent and personalized services.

### 6.3.1 *Comfort Enhancements Brought by Brain–Computer Interfaces to Transportation Tools*

Travel comfort is one of the key focuses in the design of transportation vehicles. From the initial improvements in hardware comfort to the current intelligent services, the development of transportation technology is continuously optimizing the public's travel experience. The core of BCI technology lies in monitoring human brainwaves, which provides a clearer direction for enhancing the comfort of transportation vehicles in terms of details and personalized needs.

From the application perspective of BCI technology, this technology can offer unique advantages in seat adjustment, environmental control, shock absorption, and noise reduction.

**1. Dynamic Seat Adjustment**

Comfortable seats and correct posture are especially important for long-duration travel. BCI technology combined with the control systems of transportation vehicles can automatically adjust the position and angle of the seat based on the user's real-time status, providing the most comfortable sitting posture. For example, the system can monitor the user's brainwaves, identify their fatigue state, and automatically adjust the seat angle to improve blood circulation and reduce fatigue.

In addition, the BCI system can create personalized seating settings based on the user's body shape and personal preferences, and automatically adjust to the optimal state each time the vehicle starts, enhancing comfort. For instance, some luxury car manufacturers are researching the use of BCI technology to achieve automatic seat adjustment to meet different users' needs. After the driver enters the vehicle, the car can automatically adjust the seat according to the driver's height, driving habits, and driving status, thereby providing a more comfortable driving experience.

**2. Environmental Control**

Environmental control inside transportation vehicles is crucial to the user's comfort experience. BCI technology can monitor the user's brain signals and adjust environmental parameters such as temperature and humidity in real time. Similarly, the BCI system can adjust lighting brightness and audio systems according to the user's state and preferences. For example, during long-distance driving, when the user feels fatigued, the system can switch the lighting to a soothing mode and play soft music to help the user relax. Such intelligent adjustments not only improve comfort but also effectively enhance the user's mood and psychological state.

**3. Shock Absorption and Noise Reduction**

A smooth and quiet ride experience is an important component of comfort. BCI technology can monitor the user's reactions to shock absorption and noise in real time and automatically optimize the suspension system and noise reduction levels. For example, during vehicle operation, the system can dynamically adjust the hardness and damping coefficient of the suspension based on road conditions and user feedback to provide a smooth ride. In terms of noise control, BCI technology can also automatically adjust noise reduction system settings by monitoring the user's response to noise, reducing noise levels inside cars, ship cabins, or airplane cabins.

It is evident that BCI technology enables transportation vehicles to provide more considerate and detailed driving and riding services, and this technology will be deeply applied in future transportation vehicles.

### 6.3.2 *Convenience Improvement Brought by Brain–Computer Interfaces to Transportation Vehicles*

In addition to enhancing the comfort of vehicles, BCI technology can also bring more convenience to transportation. Especially when combined with transportation technology, the value of BCI technology can be fully highlighted.

By analyzing the current research and application status of BCI technology in the field of transportation technology, it can be seen that this technology is currently strengthening its applications in areas such as intelligent navigation, voice (gesture) control, and information integration, with a very prominent improvement in convenience brought by these fields.

#### 1. Intelligent Navigation

The application of BCI technology in intelligent navigation and route planning enables users to more conveniently obtain personalized navigation services. Although traditional navigation systems are accurate, they mostly require text and voice input, and their personalized route settings can only be planned as a whole, not detailed to each segment. However, after integrating BCI technology, the navigation system can monitor the user's EEG signals to understand their preferences and needs in real time, providing the best route suggestions. For example, the navigation system can change route segments at any time while keeping the destination unchanged according to the user's status, ensuring the best driving or riding experience. In addition, the BCI system can also combine real-time traffic information to dynamically adjust navigation routes, ensuring that users always choose the optimal path.

#### 2. Voice and Gesture Control

Currently, the way vehicles receive commands has upgraded from text to voice, and BCI technology can further enhance the command reception by upgrading voice input to gesture or thought input. Users can complete operations such as navigation settings and phone answering through EEG signals combined with voice and gesture commands. For example, when driving, users only need to trigger the voice assistant via EEG signals to complete various operations without distracting their attention. Similarly, the BCI system also supports gesture control, achieving more natural and intuitive human–machine interaction by monitoring users' EEG signals and gesture movements.

**3. Information Integration**
BCI technology can also achieve seamless connection and information integration with other smart devices, improving the operational convenience of vehicles. For example, users can control devices such as smartphones and tablets through EEG signals to complete information queries, media playback, and other operations. This can greatly enhance convenience and safety during vehicle operation.

Overall, BCI technology shows great potential in improving the comfort and convenience of vehicles. As BCI technology continues to integrate with other transportation technologies, it will play an increasingly important role in the design and use of vehicles, making public travel experiences more convenient, safe, and efficient.

## 6.4 Application of Brain–Computer Interface in Drone Control

Unmanned Aerial Vehicles (UAVs) technological product was born in the early 21st century. Initially, it was mainly used for military reconnaissance. However, with the development of technology, UAVs have gradually expanded from military to civilian applications. Today, UAVs have become important tools in fields such as power inspection, public security patrols, and traffic management, and they have also begun to play a crucial role in the construction of smart cities.[5]

In today's rapid development of UAV technology, BCI technology has also begun to demonstrate enormous application potential and value in this field. The combination of these two technologies breaks through the limitations of traditional remote control and automated operation, greatly expanding the application scope and effectiveness of this technological product.

[5]An unmanned aerial vehicle (UAV), commonly known as a drone, is an aircraft that operates without a human pilot onboard. It is controlled either remotely via radio equipment and its pre-programmed control systems, or autonomously through onboard computer systems that manage its flight either fully or intermittently.

As is well known, UAV operation is somewhat difficult and requires professional training to fully realize the value of UAVs. For UAVs above a certain size, Chinese law also requires obtaining the corresponding license to operate them. However, BCI technology can overcome these operational difficulties and experience limitations by receiving, computing, and processing brain signals to achieve remote UAV control. This control method not only reduces reliance on traditional remote control devices but also significantly improves operational flexibility and accuracy.

Moreover, BCI technology also demonstrates strong advantages in the intelligent planning and execution of UAV missions. By integrating artificial intelligence algorithms and big data analysis, BCI systems can automatically generate optimal mission execution plans based on the operator's task intentions and environmental data. BCI technology not only enables autonomous mission planning for a single UAV but also facilitates coordinated operation of multiple UAVs through EEG signal monitoring and data integration. This intelligent and autonomous mission planning and execution capability opens new possibilities for UAV applications in search and rescue, agriculture, environmental monitoring, and other fields.

In summary, the application of BCI technology in the UAV field has greatly expanded the control methods and application scope of UAVs, significantly enhancing their intelligence level and mission execution efficiency. This technology can generate tremendous value in the UAV domain, creating a synergistic effect where the combination of these two cutting-edge technologies results in more than the sum of their parts.

## 6.5 Potential Advantages of Brain–Computer Interface in Vehicle, Ship, and Aircraft Control

The introduction of BCI technology enables transportation control systems to obtain real-time physiological and psychological states of operators, thereby dynamically adjusting control strategies and optimizing the operational experience. Through deep integration with big data and artificial intelligence, BCI systems can achieve efficient data integration and analysis, predict and prevent potential operational risks, and provide intelligent decision support. Additionally, BCI technology can adaptively adjust operational parameters based on the operator's real-time status and environmental changes, enhancing the adaptability and emergency response capabilities of transportation vehicles in complex environments.

It is evident that after integrating with transportation technology, BCI technology fully demonstrates its value. Its potential advantages in transportation vehicle control cover multiple aspects, from data analysis and environmental adaptation to user experience optimization. The application of this technology can bring about a transformation in transportation control methods, providing users with a safer, more efficient, and more convenient travel experience.

### 6.5.1 *Advanced Data Analysis Advantages Brought by Brain–Computer Interface*

In the control of transportation vehicles, the introduction of BCI technology has brought unprecedented advantages to the development of transportation technology. This technology not only demonstrates its own value but also enables deeper application of many cutting-edge technologies, playing a greater role in the transportation field.

In the integration of BCI technology with transportation science and technology, its primary advantage is advanced data analysis. BCI technology combined with big data and artificial intelligence optimizes control strategies and decision-making processes in intelligent transportation management systems, thereby significantly enhancing operational safety and efficiency. For example, BCI technology obtains physiological and psychological state data of operators by real-time monitoring of their EEG signals. These data can be integrated with the operational data of vehicles, ships, or aircraft as well as environmental data to form a comprehensive analysis platform.

Additionally, combined with artificial intelligence technology, BCI systems can perform predictive maintenance by analyzing historical data and real-time monitoring data to forecast potential failures and carry out maintenance in advance. This not only improves the reliability of transportation vehicles but also reduces the occurrence of unexpected incidents. For example, current hardware monitoring methods for transportation vehicles mostly analyze real-time hardware data, but after integrating BCI technology, the management system of the vehicle can also incorporate the experience of drivers and operators. When drivers or operators have concerns about certain hardware or software functions of the vehicle based on their own experience, the system can focus on monitoring these hardware components or functional modules or perform early maintenance, effectively ensuring the safety and stability of the vehicle.

Second, BCI technology demonstrates unique advantages in personalized driving and operation prediction. By continuously monitoring the operator's EEG signals, the BCI system summarizes the driver's operational habits and predicts their operational intentions and behaviors when certain specified actions or emotions occur, optimizing the driving and operational experience and enabling proactive driving assistance and intervention. For example, when the system predicts that the driver is about to perform a lane change, it can carry out route planning and safety checks in advance, improving the smoothness and safety of the operation.

The advanced data analysis advantages of BCI technology in vehicle control provide strong support for optimizing control strategies, enhancing operational safety, and improving efficiency. Through real-time monitoring, data-driven decision-making, intelligent predictive maintenance, and personalized driving prediction, BCI technology not only increases the reliability and operational efficiency of vehicles but also opens new pathways for the intelligent development of transportation tools. In the future, with continuous advancements and wider application of BCI technology, vehicle control methods will undergo new transformations, bringing users a more intelligent and personalized travel experience.

## 6.5.2 *Environmental Adaptation Advantages Brought by Brain–Computer Interfaces*

In addition to the advantages brought by advanced data analysis, BCI technology also enhances the environmental adaptability of transportation vehicles. The introduction of BCI technology enables vehicles to operate efficiently in complex and variable environments. This is mainly reflected in its rapid adaptation capability to dynamic environments.

**1. Operation in Complex Environments**

When facing harsh weather or complex terrain conditions, the BCI system can dynamically adjust control strategies by real-time monitoring of the operator's EEG signals combined with environmental sensor data. For example, under severe weather conditions such as heavy rain or snow, the driver's attention and reaction time are affected, and drivers usually cannot consider all environmental factors. At this time, BCI technology can intervene in driving based on the driver's actual state. For instance, it can automatically adjust the driving mode to ensure the safety status of the vehicle under adverse conditions.

**2. Adaptive Control**

BCI technology can not only monitor environmental changes in real time but also autonomously adjust control parameters to improve the vehicle's environmental adaptability. The most common application scenario is urban traffic, where the BCI system can dynamically adjust the vehicle's driving mode and route based on real-time traffic conditions and the driver's state to avoid congestion and improve driving efficiency. Similarly, in marine environments, the BCI system can automatically adjust navigation strategies according to sea conditions and the captain's status to ensure safe navigation in complex waters.

**3. Emergency Decision Support**

In emergency situations, the BCI system can quickly make decisions by real-time monitoring the operator's EEG signals. For example, when an aircraft encounters sudden turbulence or mechanical failure, the BCI system can rapidly adjust flight parameters, such as changing altitude or speed, based on the pilot's condition and real-time environmental data to prevent accidents.

**4. Multimodal Environmental Perception**

BCI technology can also integrate data from various environmental sensors, such as radar, LiDAR, and cameras, to form a multimodal perception system. This multimodal perception system can provide more comprehensive and accurate environmental information, helping vehicles better understand and respond to complex environments.

This aspect of BCI technology is crucial for the development of autonomous driving technology. In autonomous vehicles, traditional autonomous driving technology can combine environmental sensor data to provide real-time environment monitoring and obstacle detection, thereby making autonomous driving decisions. However, with the aid of BCI technology, autonomous driving can also incorporate the driver's driving experience, optimizing driving decisions from the perspective of drivers and passengers, and enhancing the riding experience.

Overall, BCI technology enables vehicles to fully adapt to various environments, not only ensuring vehicle safety but also providing further promotion and optimization for autonomous driving and intelligent assistance.

### 6.5.3 *Brain–Computer Interface Promotes Brain-Controlled Operation of Drones*

Although practical applications have not yet been realized, countries worldwide have made EEG-based control of various intelligent devices a major research focus, and in the field of drones, this research has already entered a substantive stage.

**MetaBOC (BOC stands for brain-on-chip):** In July 2024, the research team from Tianjin University announced their research achievement, MetaBOC (BOC stands for brain-on-chip). MetaBOC was collaboratively developed by the Haihe Laboratory of Brain–Computer Interaction and Human–Machine Integration team at Tianjin University and Southern University of Science and Technology. It is the world's first open-source on-chip brain intelligent composite information interaction system. This research achievement opens a new path for BCI technology and has brought revolutionary advancements to cutting-edge fields such as hybrid intelligence and brain-like computing. Figure 6.2 is a schematic diagram.

Nowadays, this human brain-like organ connected to a computer chip can control robots to perform various actions such as obstacle avoidance, tracking, and grasping, accomplishing pioneering work in multiple types of brain computation. Some research results have already been published in the international brain science journal *Brain*. As early as 2018, Shenzhen Chuangda Yunrui Intelligent Technology Co., Ltd. in China successfully independently developed an innovative drone interaction system integrating BCI technology, motion sensing technology, and computer vision technology. Based on this, they created a thought-controlled headband for drone operation. The drone was named "UDrone Thought-Controlled Drone." Although the "UDrone Thought-Controlled Drone" at

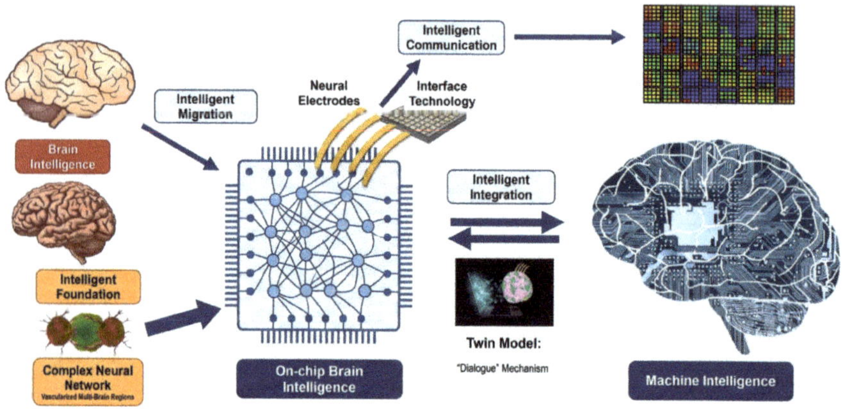


**Figure 6.2.** MetaBOC schematic diagram.

that time could not perform highly difficult or complex operations, it achieved functions such as controlling drone flight by thought, blinking to take photos, clenching teeth to land, and controlling flight direction with motion sensing actions. It is evident that using BCI technology to control drones has become an important development direction in the drone field.

Currently, the technical theory of controlling drone operations through BCI technology is very clear. BCI technology can directly control drones wirelessly. Operators can wear EEG devices to transmit brain signals to the drone control system in real time. After interpreting these signals, the system converts them into specific control commands such as takeoff, landing, and turning. This wireless brain control technology not only eliminates the dependence on traditional remote control devices but also makes operation more convenient. For example, operators can increase the drone's altitude by concentrating their attention or land the drone by relaxing their brain.

Current research results have shown that by real-time monitoring and interpreting the operator's EEG signals, BCI technology can achieve precise control of drones. Compared with traditional remote controller operations, brain control can significantly shorten the command transmission time and improve the response speed and accuracy of operations. In complex flight tasks, operators can directly control subtle drone movements through specific mental activities, such as imagining certain actions or concentrating attention. This level of operational precision is difficult to achieve with remote control. Moreover, this high-precision control method is especially important in scenarios requiring rapid response and high accuracy, such as emergency rescue and reconnaissance in hazardous areas.

Additionally, BCI technology provides drones with greater flexibility and diversity in operation. Operators can flexibly adjust the drone's operating modes through different EEG signal patterns. For example, during search and rescue missions, operators can switch the drone's flight mode from rapid search to detailed reconnaissance via brain signals. Furthermore, brain control can be combined with gesture control and voice commands to form a multimodal operation method, enhancing overall operational flexibility and efficiency.

It can be seen that the introduction of brain control enables drones to demonstrate great potential in more application scenarios. For example, in agricultural monitoring, operators can remotely control drones via brain control technology to patrol and collect data over large farmland areas; in environmental protection, operators can use brain-controlled drones for air quality monitoring and pollution source tracking; in the military field, brain-controlled drones can be used for battlefield reconnaissance and tactical deployment. The expansion of these application scenarios greatly enriches the use cases of drones and provides management and monitoring assistance across various industries, significantly promoting intelligent management of specific regions.

Various signs indicate that BCI technology will become an important trend in drone brain control operations. The value generated by the integration of these two technologies will benefit multiple industries while promoting the intelligent development of these sectors.

### 6.5.4 *Brain–Computer Interface Promotes Intelligent Planning and Execution of Drone Missions*

With the intelligent development of drones, they will undertake and execute more complex tasks, and the integration of BCI technology can realize greater value in the intelligent development of drones.

First, BCI technology combined with artificial intelligence algorithms can help drones achieve autonomous task planning and execution. This capability allows drones to demonstrate higher efficiency and intelligence in complex tasks. For example, by monitoring the operator's EEG signals, the BCI system can accurately interpret the operator's task intentions and convert them into specific task commands. In search and rescue missions, the operator only needs to indicate the drone to search a specific area through mental activity. Upon locating the target, the BCI system, combined with AI algorithms, quickly plans an escape route to help trapped or injured individuals evacuate and receive treatment in the shortest time possible.

Second, the BCI system combined with big data, artificial intelligence, and other technologies can not only execute initial task planning but also dynamically adjust and optimize task planning in real time based on monitoring the operator's status and emotions during task execution.

For example, when a drone is performing an environmental monitoring task, if the operator detects a new pollution source, the BCI system can recalculate the monitoring task data through EEG signals and adjust the task execution commands, transforming the routine monitoring plan into a targeted monitoring plan, thereby improving the task's flexibility and response speed immediately.

Finally, the introduction of BCI technology has made the coordinated operation of multiple drones possible. That is, a single operator can simultaneously control multiple drones to work together, greatly enhancing the drones' ability to perform complex tasks.

In multi-drone coordinated operations, the BCI system can automatically allocate and coordinate the tasks of each drone based on mission requirements and environmental data. For example, in agricultural monitoring, the operator can use brain-control technology to direct multiple drones to patrol and collect data in different areas. The system will automatically coordinate the flight paths and task assignments of each drone to ensure maximum coverage and efficient data collection. Similarly, in disaster emergency response, multiple drones can achieve real-time coordinated operations through the BCI system, quickly deploying and adjusting according to the operator's instructions, providing efficient disaster reconnaissance and delivery of relief supplies. This real-time coordination capability improves the drones' response speed and task execution efficiency in emergency situations.

In summary, BCI technology brings not only improvements in control precision and convenience to drone development but also an intelligent advancement in drone technology. Supported by BCI technology, drones will become multifunctional tools across various fields and industries, providing comprehensive technical support for transportation development, public life, and social progress.

# Chapter 7

# Brain–Computer Interface and Traffic Signal Control

## 7.1 Brain–Computer Interface and Traffic Signal Phase Control

In modern urban traffic management, traffic signal phase control systems are key to maintaining traffic order, reducing traffic accidents, and improving road traffic efficiency. However, despite the achievements of existing traffic signal control systems, many challenges remain.[1]

Traditional traffic signal phase control systems mainly rely on preset schedules and sensors. These systems typically control signal light switching based on fixed time plans. On busy urban roads, the switching times of traffic lights are usually fixed, aiming to adjust according to peak and off-peak traffic flows. This approach can alleviate traffic pressure to some extent but also has obvious limitations.

Another common control method is the adaptive control system, which uses sensors installed on roads to monitor vehicle flow and speed in real time and dynamically adjust the signal light phases based on this data. Compared to fixed-time control, adaptive control systems better adapt to changes in traffic flow, improving traffic continuity and throughput

---

[1] Within a signal cycle, if one or several traffic streams receive identical signal light indications at any given moment, the continuous sequence of distinct signal phases (e.g., green, yellow, and all-red) assigned to them is referred to as a signal phase.

**Figure 7.1.** Schematic of vehicle induction control.

efficiency. However, the effectiveness of these systems largely depends on the sensitivity of the sensors and the accuracy of data processing.

Vehicle induction control is also a relatively common technology, which triggers signal light switching by installing induction coils under the road surface that detect passing vehicles. This technology is widely used at intersections and smaller road crossings, effectively reducing waiting times on empty roads and improving traffic efficiency. Figure 7.1 shows a schematic of vehicle induction control.[2]

It can be seen that various traffic signal phase systems have been widely applied in traffic management. Although they demonstrate unique value under different circumstances, there are still significant shortcomings. These shortcomings and limitations are mainly reflected in the following aspects.

---

[2]Inductive signal control does not have a fixed cycle length. Its operational principle involves installing vehicle arrival detectors at the approach lanes of intersections managed by inductive signals. Inductive signal control modes are categorized into semi-actuated control, fully actuated control, and coordinated control. However, inductive signal control is rarely adopted in urban traffic management systems in China, primarily due to its high cost and the incompatibility of its configuration with the characteristics of urban road traffic composition in the country. Nonetheless, inductive control offers flexible operation and holds promising application prospects.

### 7.1.1 *Five Core Limitations of Traditional Traffic Signal Control Systems: From Response Lag to Systemic Coordination Deficits*

**1. Limited Response Speed:** Traditional traffic signal control systems are usually based on preset schedules and sensor data, lacking the ability to quickly respond to sudden traffic conditions. For example, when a traffic accident or temporary road closure occurs, existing systems may fail to adjust in time, resulting in aggravated traffic congestion.

**2. Insufficient Real-time Capability:** Although adaptive control systems can adjust based on real-time data, their response speed and accuracy depend on the performance of sensors and data processing systems. Sensors may be affected by weather and environmental factors, leading to inaccurate data, which in turn affects the system's real-time capability.

**3. Lack of Flexibility:** Existing control systems often lack sufficient flexibility when facing complex and variable traffic conditions. For example, during holidays, special events, or emergencies, traffic volume and flow directions may change significantly, but fixed schedules and limited adaptive adjustment capabilities struggle to cope with these changes.

**4. Strong Data Dependency:** Many control systems heavily rely on sensors and cameras installed on roads. These devices have high installation and maintenance costs and are susceptible to environmental impacts (such as sensor failures caused by rain or snow). Additionally, delays or errors may occur during data transmission and processing, further affecting system performance.

**5. Insufficient Coordination:** Within a city or region, different types of traffic signal control systems may coexist, and coordination among these systems is often difficult, resulting in low overall traffic management efficiency. Moreover, data sharing and information exchange between different systems face obstacles, affecting the optimization of traffic flow.

Facing the above challenges, the introduction of brain–computer interface (BCI) technology brings multifaceted improvements to traffic signal phase control systems, enabling these systems to exhibit higher intelligence. In recent years, BCI technology has made significant progress in fields such as healthcare, education, and entertainment, and its application in traffic management has gradually become a research hotspot. Especially in the aspect of traffic signal phase control, BCI technology has demonstrated enormous potential. By combining the characteristics and principles of BCI technology, we can identify many conveniences it brings to traffic signal phase control systems.

For example, on the main roads of a city, traffic flow varies at different times, and traffic pressure also differs. If relying solely on external data monitoring tools, traffic management departments can only respond based on monitoring data when a certain flow pressure appears on the main roads. However, after integrating the BCI system, traffic management departments can make adjustment decisions for the traffic signal phase control system in advance based on numerical changes in drivers' emotional states, thereby alleviating congestion on main roads more promptly.

Another example is in the case of traffic accidents or emergencies, where traditional signal control systems often fail to respond in a timely manner. Through BCI technology, traffic management personnel can quickly issue adjustment commands in emergency situations. For instance, when a traffic accident requires temporary closure of a certain road section, the management personnel can send commands via the BCI system, and the system immediately adjusts the signal phases of the relevant road sections to ensure the redistribution of traffic flow and rapid restoration of order.

It can be seen that the application of BCI technology in traffic signal phase control has multiple advantages, which significantly promote the resolution of urban traffic pressure and improve the public travel experience. However, although integrating BCI technology into traffic signal phase control systems is an important direction in traffic technology development, there are still some technical challenges in its practical application. Only by comprehensively solving these issues can the comprehensive design of existing traffic signal phase control systems be realized, allowing BCI technology to be successfully implemented in this field.

### 7.1.2 *Key Challenges in BCI-Integrated Traffic Signal Control: Acquisition Accuracy, Processing Speed, and System Reliability*

**1. Control Signal Acquisition Accuracy**

The acquisition of brain signals requires high-precision equipment and technology, and any interference can affect the accuracy of the signals.

This necessitates the adoption of more advanced sensor technologies and signal processing algorithms to improve the accuracy and stability of signal acquisition. Currently, this remains a major challenge in BCI technology research and development, both in hardware and software aspects.

**2. Improving Data Processing Speed**
Although we have entered the era of big data and data processing capabilities in various fields are continuously improving, integrating BCI technology into traffic signal phase control systems requires extremely high real-time signal acquisition and data processing capabilities. Existing signal acquisition and data processing technologies cannot perfectly meet the integration demands of BCI technology, which is a significant difficulty in incorporating BCI into traffic signal phase control systems.

**3. Enhancing System Stability and Reliability**
The stability and reliability of BCI systems are crucial, as any instability or malfunction may lead to chaos in traffic management. Therefore, integrating this technology into traffic signal phase control systems requires the establishment of redundant systems and fault detection mechanisms to ensure stable system operation. This is currently a key research focus in the field of BCI technology.

Although there are still the aforementioned challenges in integrating BCI technology into traffic signal phase control systems, this direction is already very clear. BCI systems can improve the response speed, real-time performance, and flexibility of traffic signal control, optimize traffic flow, reduce congestion, and enhance traffic efficiency. It is believed that through continuous innovation and technological advancement, BCI technology will soon make significant contributions to building smarter, more efficient, and safer traffic systems.

## 7.2 Brain–Computer Interface and Traffic Signal Priority Control

The traffic signal priority control system is an important tool to address urban traffic accidents and ensure the smooth overall operation of city traffic. Its main goal is to guarantee that specific vehicles such as public transportation and emergency vehicles can pass through intersections with priority, thereby improving traffic efficiency, reducing delays, and ensuring the timely completion of emergency tasks.

Currently, in some cities of developed countries abroad, traffic signal priority control systems operate based on preset traffic rules and hardware sensing devices installed on urban roads. However, this control method is still in the experimental stage. For example, the Transit Signal Priority (TSP) system and Emergency Vehicle Priority (EVP) system are two common priority control methods. The TSP system usually achieves priority passage for buses through communication between transmitters installed on buses and traffic signal controllers, as shown in Figure 7.2. The EVP system, on the other hand, uses special equipment on emergency vehicles (such as sirens) to trigger priority access when approaching intersections.

These systems and devices have improved traffic efficiency to some extent and reduced delays for public transportation and emergency vehicles.

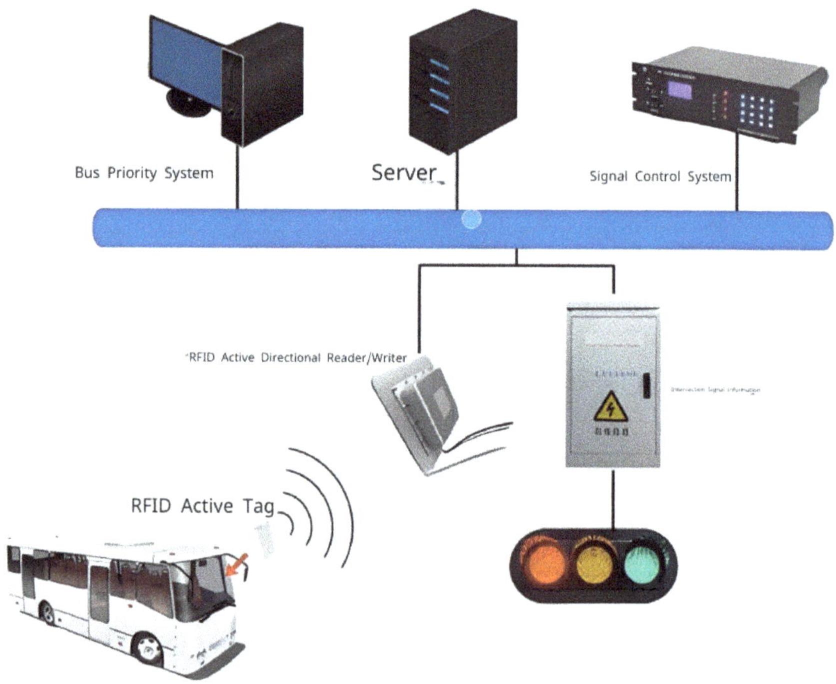


**Figure 7.2.** Transit signal priority system.

However, traditional systems and devices still have some limitations, so many problems remain in actual operation. For example, as early as 2017, in Jinan City, China, there was an incident where private cars not only failed to yield priority to an ambulance but even deliberately blocked it, resulting in the patient not receiving timely treatment and ultimately dying. Such situations occur annually across the country.

Overall, in modern urban traffic, traffic flow is dense and highly variable, and different types of vehicles have different demands for traffic signal priority control. For instance, buses need priority passage during peak hours to ensure punctual public transportation; emergency vehicles such as ambulances and fire trucks require rapid crossing of intersections at any time to reduce emergency response time. Therefore, this places higher demands on traffic signal priority control. However, the current intelligent control systems for traffic signals in China still have some shortcomings, which precisely calls for cutting-edge technology to improve traffic signal priority control, and BCI technology has become the best choice.

Through BCI technology, traffic management personnel or vehicle drivers can adjust the priority of traffic signals in real time. For example, in emergencies, an ambulance driver can send a priority passage command via the BCI system. Upon receiving the command, the system immediately adjusts the signal phases at relevant intersections based on the vehicle's movement, ensuring priority passage for emergency vehicles. This maximizes the road priority rights for special vehicles while enhancing the response speed in emergencies.

From the application of BCI technology in traffic signal priority control, we can see that this technology can comprehensively upgrade the traffic signal priority control system. The changes it brings mainly manifest in the following two aspects.

**1. Improving the flexibility of system control**

By capturing and interpreting brain signals in real time, BCI technology can quickly identify and execute priority control commands, enhancing the system's real-time responsiveness and flexibility. Compared with traditional traffic signal priority control systems, the BCI system is more adaptable to sudden traffic conditions, capable of timely adjusting traffic signals to ensure smooth and efficient traffic flow.

**2. Enhancing the Coordinated Control Effect of Multiple Traffic Lights**

Under normal circumstances, when any emergency occurs, special vehicles require not only priority control assistance at a single intersection's traffic light but also the overall priority right of way along an entire route. This necessitates the coordinated control of multiple traffic lights. BCI technology can precisely achieve seamless integration of multiple intersection traffic lights by sharing real-time data and exchanging information, enabling overall control of the traffic lights along a route to ensure that special vehicles maintain smooth passage throughout their mission.

In the future, with the further development of BCI technology, we can foresee its broad application in the field of traffic signal control as well as the entire intelligent transportation management domain. Its unique value and role will contribute significantly to building a smarter, more efficient, and safer traffic system, creating more possibilities for the development of intelligent transportation in our country.

## 7.3 Brain–Computer Interface and Adaptive Control of Traffic Lights

The traffic light system, as the nerve center of urban traffic control, directly affects the city's traffic conditions and the quality of life of its residents through its design and operation. With technological progress and social development, traffic light systems have continuously evolved—from initially simple mechanical operations to today's highly complex adaptive control systems. Each transformation aims to improve traffic efficiency and safety.

Traditional traffic light systems are based on mechanical and electronic technologies, primarily relying on timers and simple vehicle detection techniques to control the switching of red and green lights. These systems are usually designed based on historical traffic flow data, setting different red-green light switching schemes for various times of the day. For example, during morning and evening peak hours, the traffic light cycles are adjusted to reduce waiting times and increase traffic throughput.

However, this fixed-mode operation has significant limitations. First, it cannot adapt to emergencies or irregular situations, such as abnormal traffic flow caused by large events or accidents. Second, traditional systems fail to effectively integrate traffic lights at adjacent intersections, resulting in the so-called "island effect," where each intersection's traffic lights operate independently without effective coordination, reducing overall traffic efficiency. Additionally, these systems often do not consider the special needs of non-motorized vehicles such as pedestrians and bicycles.

With the acceleration of urbanization and the sharp increase in private vehicles, traditional traffic light systems face unprecedented challenges. Traffic congestion not only increases commuting time but also exacerbates environmental pollution. Studies show that approximately 10% to 20% of urban traffic congestion is caused by unreasonable traffic light control. Moreover, congestion leads to frequent stopping and starting of vehicles, increasing fuel consumption and exhaust emissions.

To address the limitations of traditional systems, adaptive traffic signal control systems have emerged and are gradually being promoted worldwide. Adaptive control systems use advanced sensing technologies, such as ground sensors and video surveillance, to collect real-time traffic flow data and analyze it through a central processing unit, dynamically adjusting the operating status of traffic lights.

The core of this system lies in its adaptive algorithm, which can adjust the duration, sequence, and frequency of traffic lights based on real-time data to optimize traffic flow. For example, if an increase in vehicles is detected in a certain direction, the system will extend the green light time for that direction while relatively reducing green light time for other directions. In adaptive control systems, traffic signals no longer switch simply according to fixed patterns. Instead, control authority is transferred to an intelligent system that can analyze multiple factors such as vehicle speed, traffic volume, and pedestrian numbers to make the best decisions accordingly.

For example, the traffic signal control system in New York City, USA, can adjust signal changes based on real-time analysis of traffic data according to actual demand, reducing traffic delays by approximately 10%. At the same time, the system can automatically recognize the alarms of special vehicles, such as fire trucks and ambulances, and open green channels for them.

Additionally, the adaptive traffic signal system can be integrated with other intelligent systems such as urban traffic navigation systems and public transportation systems, forming a comprehensive smart city traffic network. Through big data analysis and artificial intelligence algorithms, more refined traffic control and prediction can be achieved.

Nowadays, with the integration of BCI technology and transportation technology, the adaptive traffic signal system has gained significant improvement potential. This integration not only greatly enhances the responsiveness and intelligence level of the traffic system but also provides possibilities for realizing more personalized traffic services.

The introduction of BCI technology can bring disruptive changes to the traditional adaptive control system of traffic signals.

First, through high-precision electroencephalography (EEG) capturing devices, such as wearable sensors or sensing devices embedded in safety helmets and hard hats, the system can monitor in real time special signals that require priority passage, such as the current emergency status of drivers of ambulances or fire trucks. The system analyzes these brainwaves and automatically adjusts the traffic lights to ensure these vehicles can pass quickly.

Moreover, for pedestrians, especially those with mobility impairments, BCI technology can also play a role. By recognizing specific EEG patterns, the system can determine whether a pedestrian intends to cross the street and appropriately extend the green light duration at crosswalks to ensure their safe crossing.

Finally, the integration of BCI technology directly improves the flexibility and efficiency of the traffic system. On one hand, by reducing unnecessary stops that cause exhaust emissions, it benefits the improvement of the urban environment; on the other hand, it enhances emergency response speed and strengthens public safety.

Therefore, the integration of BCI technology with adaptive traffic signal control systems presents a new approach to solving urban traffic problems. With the maturity of related technologies and the reduction of costs, this integrated technology can be promoted and applied in more cities in the future, contributing to the construction of smarter, more efficient, and more human-centered traffic systems.

# Chapter 8

# Brain–Computer Interface and Traffic Information Processing

## 8.1 The Role of Brain–Computer Interface in Traffic Information Recognition

With the acceleration of urbanization, urban traffic faces increasingly severe challenges such as congestion, frequent traffic accidents, and environmental pollution. To address these issues, traffic information recognition technology has emerged and quickly become an important tool for improving traffic efficiency and ensuring traffic safety.

Traffic information recognition technology provides scientific decision-making support for traffic management departments by real-time monitoring and analyzing traffic flow, vehicle speed, traffic density, and other information on the roads. These technologies not only effectively alleviate traffic congestion and improve road traffic efficiency but also reduce the incidence of traffic accidents and enhance road safety levels. At the same time, traffic information recognition technology plays a positive role in environmental protection by optimizing traffic flow, thereby reducing exhaust emissions and improving air quality. Additionally, this technology is an important tool for law enforcement agencies in China to solve cases, playing a crucial role in ensuring public safety.

Due to the significant importance of traffic information recognition technology, its development has always been a key focus in China's transportation sector. In recent years, this technology has made remarkable progress in both hardware and software aspects. Compared to the period before 2020, traffic information recognition technology has achieved outstanding results in the following areas.

1. **Significant Improvement in Data Collection Accuracy**
Before 2020, China's traffic information recognition technology mainly relied on traditional monitoring equipment. However, after 2020, with the rapid development of sensor technology, the transportation departments in China have been able to obtain higher-precision and more real-time traffic data. The collection of these high-precision data mainly depends on more advanced intelligent cameras, LiDAR, and high-precision GPS devices.

2. **Significant Increase in Processing Speed**
With the advent of the big data era, the massive growth of traffic data and the development of intelligent transportation have faced many challenges. Precisely identifying target information within this larger volume of data requires faster processing speeds and more efficient algorithms. However, since China's traffic management departments have integrated cutting-edge technologies such as cloud computing and AI, data processing capabilities have multiplied, and significant progress has been made in traffic information recognition capabilities.

3. **Enhanced Environmental Adaptability**
In the past, traditional traffic information recognition facilities were easily affected by environmental factors, causing fluctuations in equipment lifespan, data collection accuracy, and recognition capabilities. However, with the development of transportation technology, these devices have undergone a qualitative leap in recent years. Not only has the accuracy of data collection improved, but environmental adaptability has also been strengthened, making traffic information recognition systems more stable and precise. Taking China's well-known traffic violation monitoring system—the Haiyan System—as an example, it can currently recognize over 1,000 photos per minute, accurately monitoring any violation, with this monitoring capability being minimally affected by environmental conditions.

4. **Improved Data Fusion Capability**
Traffic information recognition technology places great emphasis on the fusion of multi-source data by integrating data from different sensors, cameras, and user devices to provide more comprehensive and accurate

traffic information. This includes the fusion of various perceptual data such as visual, auditory, and tactile inputs, enhancing the system's overall recognition capability. Currently, the scope of data fusion in China's traffic recognition systems has been comprehensively expanded—from initially focusing on image data related to violations to including image and sound data, and even integrating environmental monitoring data in some regions. This improvement in data fusion capability has brought more assistance to traffic management, especially in traffic planning and emergency response, allowing traffic departments to make more precise decisions based on data recognition. For example, in some regions of China, traffic departments have cooperated with environmental monitoring agencies to integrate environmental monitoring data into traffic information recognition systems. Traffic management departments can also promptly identify environmental conditions and control quality data, implementing corresponding "license plate restrictions" and "traffic restrictions" measures.

**5. Extensive Application of Artificial Intelligence**

Artificial intelligence technology has now become an important driving force for traffic information recognition technology. Currently, traffic management departments across the country are striving to build their own intelligent traffic management systems. By applying advanced technologies such as deep learning and neural networks, these systems will be able to learn and improve autonomously, enhancing the accuracy and efficiency of traffic information recognition.

From the above five points, it can be seen that the development of traffic information recognition capabilities in China is moving toward technological advancement, high-end development, and intelligence. In this process, brain–computer interface (BCI) technology has also begun to demonstrate its value and shows outstanding advantages in traffic recognition capabilities.

The core of BCI technology lies in its ability to directly convert the brain's intentions and thoughts into operational commands, thereby achieving precise control over external systems. The application of BCI technology has already achieved significant results in fields such as medical care and entertainment. In the field of traffic information recognition, BCI technology also shows great potential. Especially regarding the five development directions mentioned above for traffic information recognition, BCI technology can bring comprehensive upgrades.

First, in terms of the accuracy of information collection. BCI technology achieves highly real-time and precise traffic information recognition by directly reading and interpreting brain signals. Compared with traditional traffic information recognition technologies, BCI systems can respond more quickly to changes in traffic conditions, providing more timely and accurate information. For example, in the event of a traffic accident, the BCI system can immediately identify the severity and urgency of the accident through the brain signals of the driver or passengers, enabling the traffic emergency department to deploy more appropriate handling methods rather than conducting targeted dispatch after arriving at the scene.

Second, BCI technology has powerful multi-source data fusion capabilities. The brain signals it collects include various sensory data, such as visual, auditory, and tactile information. For instance, the BCI system can recognize traffic conditions or road changes through the driver's feelings of doubt, hesitation, or anxiety. This fusion of multi-source data can enhance the comprehensive recognition ability of traffic management systems and improve traffic management effectiveness.

Third, BCI technology can deeply integrate with AI technology and enhance traffic intelligence. When intelligent traffic management systems incorporate BCI technology, they can deeply understand urban traffic demands based on the driver's brain signals, thereby optimizing traffic development directions and making traffic development better aligned with the actual needs of the public.

Finally, BCI technology can operate stably under various complex environmental conditions. Compared with traditional traffic information recognition technologies, BCI systems are less affected by external environmental factors such as weather and lighting. For example, under harsh weather conditions such as heavy rain, snow, or dense fog, traditional cameras and sensors are somewhat affected, whereas BCI systems recognize traffic information through brain signals, maintaining stable performance and high recognition efficiency. This ensures the effectiveness of traffic information recognition systems under extreme conditions.

Currently, the application and innovation of BCI technology in the field of traffic information recognition are ongoing, and the depth of its application is continuously increasing. Through the continuous deep application of this technology, the traffic departments' traffic information recognition capabilities will significantly improve. Whether in optimizing traffic flow, enhancing road throughput efficiency, or improving the intelligence level of traffic management, BCI technology will bring remarkable enhancement effects, promoting the healthy development of cities and society with more accurate and real-time traffic information recognition capabilities.

## 8.2 Application of Brain–Computer Interface in Navigation Systems

The history of navigation systems can be traced back to ancient geography and maritime technologies. In the era before modern technology, people relied on stars, landmarks, and maps for navigation. The earliest maps were usually hand-drawn and had many limitations, such as insufficient accuracy and slow updates. Over time, navigation technology has undergone significant development from mechanical instruments to electronic devices.

At the beginning of the 20th century, with the development of radio technology, early radio navigation systems emerged. These systems were mainly used in the fields of aviation and maritime, helping pilots and captains determine their positions through radio signals. Entering the 21st century, the emergence of the Global Positioning System (GPS) completely changed the field of navigation. GPS technology uses satellite signals to provide real-time location information, making navigation more accurate and reliable. Today, modern navigation systems are not only limited to GPS but also integrate Internet data, real-time traffic information, and artificial intelligence technology to provide users with comprehensive navigation solutions.[1]

Currently, China's main navigation system is the BeiDou Navigation Satellite System, with specific applications including Baidu Navigation, Amap Navigation, as well as Google Maps and Apple Maps. These software applications utilize GPS, Geographic Information Systems (GIS), and real-time data to provide navigation services, offering detailed maps, route planning, and traffic conditions. Users can access these services

---

[1]The GPS is a high-precision radio navigation and positioning system developed and launched by the United States, based on a constellation of artificial Earth satellites. GPS is capable of providing accurate geographical location, vehicle speed, and precise time information anywhere on Earth and in near-Earth space.

through various means, such as smartphones and in-vehicle devices. At the same time, these applications can generate data based on the navigation routes used by users, providing real-time traffic information and road alerts.

Although modern navigation systems are powerful, there are still some common problems and limitations. For example, in densely populated urban areas or tunnels, GPS signals may be blocked, leading to reduced positioning accuracy. In addition, the real-time performance and intelligence level of existing systems need improvement, especially in handling complex traffic situations and personalized needs, where current navigation systems still have certain shortcomings.

At present, BCI technology, as an emerging technology, is bringing innovative applications to navigation systems and compensating for various deficiencies in navigation technology development. This technology enables seamless communication between humans and computer systems by directly interpreting brain signals. In navigation systems, it can enhance intelligence and humanization, providing users with more accurate and personalized navigation services.

First, BCI technology enables navigation systems to offer more real-time route planning. Because BCI technology can interpret users' intentions in real time by analyzing brainwave signals to understand their needs, it can dynamically adjust navigation routes. When users feel anxious or fatigued during driving, the system can automatically optimize the route, selecting safer and more comfortable paths.

Second, BCI technology can also provide more precise navigation information. It can analyze users' actual needs and adjust navigation settings in real time according to personalized requirements. For example, the system can analyze users' brainwave patterns to understand their preferences for specific routes, transportation modes, or driving styles, thereby offering personalized navigation services.

Finally, BCI technology can seamlessly connect with existing navigation systems, enhancing their performance through data fusion and intelligent algorithms. For instance, the BCI system can combine with GPS, sensors, and real-time traffic data to provide more comprehensive and intelligent navigation services. Moreover, BCI technology can collaborate with in-vehicle systems, smartphones, and other devices to achieve a multi-platform navigation experience.

Therefore, the development of BCI technology in the navigation field is not only a trend but also a key to the comprehensive upgrade of modern navigation technology. We can imagine that once BCI technology is integrated with navigation technology, a complete upgrade of navigation software can be achieved merely from the operational level. For example, blind or visually impaired individuals can easily obtain accurate navigation through this technology. Using BCI technology, navigation systems can monitor the brain signals of blind users in real time and provide navigation guidance based on their cognitive state and environmental information. In practical applications, BCI technology can help visually impaired individuals better perceive their surroundings, improving the safety and convenience of pedestrian navigation.

From a technical perspective, the integration of BCI systems with navigation systems is not complicated. It only requires adding brainwave signal acquisition devices to the front end of existing navigation systems and incorporating brainwave signal analysis systems in the data analysis stage. By adding electroencephalogram (EEG) devices, signal amplifiers, sensors, and computing equipment, current navigation systems can be upgraded to BCI navigation systems.

EEG devices are used to collect users' brainwave signals, signal amplifiers enhance the signal strength, sensors collect environmental and location information, and computing devices process and analyze the data. Once the signals are accurately collected and uploaded, signal processing algorithms can filter and interpret the brainwave signals, route planning algorithms calculate the optimal navigation route, and then navigation information is generated to interact with the user.

In this system, the data processing module is the core part of the system. It includes signal preprocessing, feature extraction, pattern recognition, and decision output. Signal preprocessing is used to remove noise and interference; feature extraction is used to obtain useful information from EEG signals; pattern recognition is used to analyze and identify the user's intentions; decision output is used to generate navigation commands. This is the key and challenging focus for integrating current BCI technology into navigation systems.

Shanghai University's "Brain–Computer Interface-Based Vehicle Navigation and Control Technology" Project

Currently, Shanghai Jiao Tong University in China is conducting research on "Brain–Computer Interface-Based Vehicle Navigation and Control Technology." This research is led by Liqing Zhang as the project leader and is a major research initiative at Shanghai Jiao Tong University.

This project aims to use BCI technology to reveal the patterns of specific brain cognitive activities and establish the relationship between cognitive intentions and behaviors. This research holds significant theoretical importance for understanding the neural mechanisms of brain cognition and is valuable for developing new information perception and pattern recognition technologies capable of processing highly complex data. Additionally, the project has important application prospects in exploring human cognitive potential, developing assistive systems for disabled and elderly individuals, and particularly in the rehabilitation of cognitive impairment disorders.

The research objectives of the project include developing a novel brain–computer interaction paradigm based on scenario cognitive induction, recognition of specific cognitive evoked potential patterns, and core vehicle EEG navigation technologies, aiming to reveal the neural mechanisms of a new asynchronous and autonomous brain–computer interaction. The research content involves understanding the dynamic characteristics of EEG signal patterns induced by specific multi-modal cognitive activities, establishing characterization models and feature extraction algorithms for these evoked potentials, developing multi-modal cognitive evoked potential pattern recognition technologies, and further constructing the mapping relationship among multi-modal cognitive activities, feature representation, and decision control, namely the multi-modal brain–computer interaction paradigm.

During the research process, the team conducted in-depth explorations in three major areas: neural mechanisms of multi-task dynamic brain–computer interaction, brain signal feature extraction and pattern recognition, and typical applications of brain–computer interaction.

First, the research team proposed a bidirectional adaptive training and neurofeedback interaction model to help patients adapt more quickly to brain–computer interaction rehabilitation training systems. This training model not only improves patients' rehabilitation efficiency but also provides a theoretical foundation for multi-task BCIs. Additionally, the team designed a brain–computer interaction learning paradigm that enables the system to rapidly adapt to patients' motor imagery synchronization patterns. This innovation greatly enhances the adaptability and flexibility of BCI systems.

In terms of brain signal feature extraction and pattern recognition, the research team proposed tensor decomposition-based feature analysis and specific cognitive pattern recognition methods for high-dimensional brain signal processing. They developed non-negative tensor decomposition algorithms and tensor feature extraction algorithms for EEG time–frequency–space data. These technologies provide strong support for high-precision brain signal processing and recognition.

In practical applications, the research team successfully developed a BCI-based wheelchair control system and rehabilitation system. Clinical validation has demonstrated the effectiveness of these systems in improving mobility and rehabilitation outcomes for disabled and elderly individuals.

The development of the wheelchair control system enables disabled individuals to directly control the movement of the wheelchair through EEG signals, achieving true autonomous mobility. This system not only enhances the freedom of movement for disabled persons but also boosts their self-esteem and independence. Research shows that controlling wheelchairs via BCI technology allows disabled individuals to respond more flexibly to complex environments during travel, improving both travel efficiency and safety.

The rehabilitation system monitors patients' motor activities through EEG signals to assist in rehabilitation training. This system has demonstrated significant rehabilitation effects in clinical validation at hospitals, helping many patients regain motor abilities. Especially in stroke rehabilitation, the BCI rehabilitation system offers a novel and effective rehabilitation approach, enabling patients to recover motor functions more rapidly.

The research achievements of the project are remarkable, not only making breakthroughs in theory but also demonstrating broad application prospects in practice. The team has published a total of 24 papers, among which 5 were included in SCI, and 2 were published in IEEE Transactions series journals. In addition, the research team has presented multiple papers at important international conferences and received several awards and patents.

These results not only prove the feasibility and effectiveness of BCI technology in the field of vehicle navigation and control but also provide valuable references for the future development of related technologies. Through the project research, the team has made significant technical progress and contributed importantly to promoting the popularization and development of BCI technology in practical applications.

Overall, the project "Vehicle Navigation and Control Technology Based on Brain–Computer Interface" has achieved important breakthroughs in theoretical research and demonstrated broad prospects in practical applications. With continuous technological development, BCI technology is expected to bring more convenience and safety to the travel of disabled and elderly people, making greater contributions to social inclusiveness and diversity.

This case shows that China has achieved many results in the research of BCI technology for navigation systems. This technology not only promotes the development of intelligent transportation but also benefits more special groups and improves the overall social development effect. This is an excellent performance for building a strong transportation country and a strong technology country.

## 8.3 Brain–Computer Interface and Traffic Data Analysis

In intelligent transportation systems (ITS), data analysis is a core means to optimize traffic management, improve road safety, and enhance travel efficiency. The introduction of BCI technology also brings new dimensions and breakthroughs to traffic data analysis.

It can be seen that with the intelligent development of cities and society, the number of vehicles in China continues to increase, and traditional traffic management methods have become insufficient to cope with increasingly complex traffic conditions. Traffic departments face massive amounts of data daily, and effectively analyzing and organizing this data is an important prerequisite to ensuring healthy traffic development.

Currently, traffic data mainly comes from sensors, cameras, GPS, and other devices, covering various types such as traffic flow, driving speed, accident records, and environmental conditions. This data provides valuable decision-making basis for traffic management departments. However, traditional traffic data processing methods have limitations. For example, the coverage of fixed sensors and cameras is limited, making it difficult to comprehensively reflect the traffic conditions of the entire city; delays in data transmission and processing make real-time decision-making difficult. In addition, traditional systems cannot accurately identify the state and intentions of drivers, thus failing to provide personalized traffic management services.

Against this background, the introduction of BCI technology brings new opportunities for traffic data collection and processing. BCI technology offers a novel data collection method in the traffic field. Traditional traffic data collection methods rely on external devices, while BCI technology can monitor the EEG data of traffic management personnel or drivers in real time, combined with other sensor data, to form more comprehensive traffic information. This data collection method has unique advantages, especially in terms of real-time performance and high accuracy. As mentioned earlier, drivers tend to experience fatigue after long periods of driving. By monitoring their EEG signals through BCI technology, fatigue signs can be detected promptly, warnings can be issued to the driver, and even vehicle control can be taken over if necessary. This EEG-based data collection method enables real-time traffic data acquisition and processing, optimizing traffic management within a very short time.

Moreover, BCI technology can be applied bidirectionally in traffic data processing. EEG data can also be used to monitor the state of traffic management personnel. For example, in traffic command centers, the concentration and decision-making ability of managers are crucial. BCI technology can monitor the EEG signals of managers in real time, helping identify fatigue and stress states, thereby optimizing work arrangements and ensuring efficient traffic management operations.

In summary, BCI technology, as an innovative interaction method, is gradually changing the traditional ways and capabilities of traffic data analysis. By directly reading and interpreting neural signals from the brain, BCI technology provides new tools and methods for traffic management and optimization. Next, we analyze in detail the application of BCI technology in the field of traffic data analysis from the perspectives of application strategies and key focuses.

First of all, as a novel technology for data collection, BCI technology plays a crucial role in application strategies within the field of traffic data analysis. Based on current traffic data analysis techniques, skillfully integrating BCI technology can achieve greater improvements in the simplest way. From the current application directions of BCI technology in traffic data analysis, its application strategies are mainly reflected in the following aspects.

### 8.3.1 *Integrating Brain–Computer Interface into Traffic Systems: Real-Time Monitoring, Dynamic Management, and Personalized Services*

1. **Real-time Data Collection and Analysis:** Traditional traffic data collection typically relies on sensors, cameras, and GPS devices. Although these devices can provide large amounts of data, they have certain limitations in terms of real-time capability and accuracy. BCI technology, by directly reading the EEG signals of drivers and pedestrians, enables real-time monitoring and analysis of traffic conditions. This opens an additional channel for information acquisition and analysis in traffic data analysis, with data collection and analysis through this channel being more timely, convenient, and efficient.

2. **Dynamic Traffic Management:** BCI technology can dynamically adjust traffic management strategies based on real-time data to improve traffic flow efficiency and safety. Specifically, BCI systems can monitor the emotions and behaviors of road users in real time and dynamically adjust the timing and cycles of traffic signals accordingly. During peak hours, by monitoring drivers' emotional and stress levels, the BCI system can extend green light durations to alleviate traffic congestion. Additionally, BCI technology can integrate with ITS to dispatch public transport vehicles in real time, enhancing public transit utilization.

3. **Personalized Traffic Services:** BCI technology can provide personalized traffic services based on individual users' needs and preferences. For example, BCI systems can monitor passengers' emotional states and travel demands, automatically adjusting bus stops and departure frequencies to offer more convenient and comfortable services. Furthermore, BCI technology can be applied to taxi and shared mobility services by analyzing passengers' EEG signals to provide customized travel routes and services, thereby enhancing user experience.

Currently, the above three strategies have become the main application methods of BCI technology in the field of traffic data analysis and are key focuses in the global integration of BCI and transportation technology. For example, NASA's research projects in the United States utilize BCI technology to detect error probabilities of pilots and air traffic controllers, thereby improving aviation safety. Additionally, research projects funded by the U.S. Department of Defense are exploring hands-free control of drones using BCI technology. This technology can also be applied to vehicle driving, enhancing the flexibility and responsiveness of traffic systems. From the existing results of these research projects, we can also observe the key application areas of BCI technology in traffic data analysis.

### 8.3.2 *Key Technologies and Implementation Pathways for BCI in Traffic Data Analysis*

1. **High-Precision EEG Signal Processing:** EEG signal processing is the core of BCI technology applications. To ensure data accuracy and real-time performance, BCI systems need to possess high-precision EEG signal acquisition and processing capabilities. Using advanced signal processing algorithms, such as Non-negative Tensor Factorization and Higher-Order Partial Least Squares methods, BCI systems can extract and analyze high-dimensional EEG data to identify useful features and patterns. This data provides important reference information for traffic management and optimization.

2. **Multi-modal Data Fusion:** In traffic data analysis, a single data source is often insufficient to provide comprehensive information. BCI technology integrates multi-modal data by combining EEG data with other traffic data (such as traffic flow data and environmental data) for comprehensive analysis. Through data fusion technology, BCI systems can consolidate data from different sources to form a complete picture of traffic information and utilize big data analytics to identify traffic patterns and trends, providing decision support.

3. **Intelligent Transportation System Integration:** To fully leverage the advantages of BCI technology, it must be seamlessly integrated with existing ITS. Through system integration, BCI systems can achieve real-time data transmission and processing and collaborate with traffic management systems, intelligent transportation infrastructure, and public transit systems. By optimizing algorithms and hardware, the system's processing speed and accuracy are enhanced to ensure real-time response and dynamic adjustment, thereby improving the intelligence level of traffic management.

**4. Experimental Validation and Application Promotion:** Experimental validation is an indispensable step in applying BCI technology to traffic data analysis. Through extensive experiments and practical applications, the performance and effectiveness of BCI systems are verified, continuously optimizing and improving the technology. Meanwhile, case studies and user feedback accumulate experience and data to support further promotion and application of the technology.

**5. Data Security Assurance:** With the widespread application of BCI technology in traffic data analysis, privacy protection issues must be given high attention. Strict privacy protection policies need to be established during data collection and processing to ensure the security and privacy of user data. Additionally, ethical standards for BCI technology should be developed to ensure compliance and morality in its application, safeguarding users' rights from infringement.

Overall, BCI technology has broad application prospects in the field of traffic data analysis. This technology can significantly enhance the efficiency and intelligence level of traffic management. It is believed that with the deepening application of BCI technology in traffic data analysis, it will make important contributions to the construction of smart cities and ITS.

## 8.4 Brain–Computer Interface and Vehicle-to-Vehicle Communication

In-vehicle communication systems are not simply call systems, but wireless communication technologies developed based on the development of ITS and sensor network technologies. This technology is very important for the development of intelligent transportation. In-vehicle communication systems achieve close integration between vehicles and traffic management departments, between vehicles, and between vehicles and other groups through vehicle-to-vehicle and vehicle-to-infrastructure communication. Through effective real-time information exchange, they greatly improve the safety and efficiency of the transportation system.

In ITS, Vehicle-to-Vehicle communication technology (abbreviated as V2V) becomes even more important.

First, V2V communication can significantly improve traffic safety. Through information sharing between vehicles, drivers can be informed in advance about traffic conditions, obstacles, and emergencies on the road ahead, allowing them to react promptly and reduce the likelihood of accidents. For example, when the vehicle in front brakes suddenly, the following vehicle can immediately receive this information via V2V communication and take timely measures such as deceleration or lane change to avoid chain rear-end collisions.

Second, V2V communication technology can enhance traffic efficiency. By sharing information such as vehicle speed, position, and travel direction, the V2V system can optimize traffic signal control and vehicle scheduling, reducing congestion and increasing road capacity.

Moreover, V2V communication technology is an important support for achieving autonomous driving. Autonomous vehicles need to operate efficiently and safely in complex traffic environments, which rely heavily on information exchange between vehicles. Through V2V communication, autonomous vehicles can achieve cooperative driving and intelligent scheduling, improving the safety and reliability of autonomous driving. For instance, on highways, autonomous vehicles can maintain synchronized distances and speeds via V2V communication, forming intelligent platoons that reduce traffic accidents.

However, as of 2024, the vehicle communication system has not yet been fully established within China's intelligent transportation framework. This is not only due to the limitations of the vehicle communication technology itself but also directly related to existing automotive R&D technologies. Nevertheless, the integration of BCI technology can effectively improve this situation. Applying BCI technology to V2V communication can enhance the real-time performance and accuracy of communication, as well as enable personalized traffic services, thereby accelerating the comprehensive establishment of vehicle communication systems.

From the application perspective of BCI technology, BCI can monitor the driver's status in real time and share this information with other vehicles. Through EEG acquisition devices, the BCI system can detect the driver's fatigue level, attention concentration, and emergency response status in real time. When a single vehicle encounters driving issues, nearby vehicles can receive relevant alerts immediately and take preventive measures in advance.

Today, with the rapid growth of vehicle ownership in China, the traffic flow on urban and intercity roads continues to increase. We can regard the vehicles on each road segment as a platoon. Imagine how much traffic safety could be improved if real-time vehicle communication systems were deployed across all road sections.

Imagine when driving on a highway, a vehicle equipped with BCI-integrated V2V communication detects the driver's emergency response ahead; at this moment, our vehicle and the vehicles behind immediately receive the relevant alert, giving us more time and space to react. Alternatively, when our vehicle has autonomous driving assistance, it can intervene promptly, which is sufficient to reduce many major traffic accidents. Or, in urban traffic environments, when a vehicle at an intersection detects the driver's emergency avoidance intention, if the intersection area is equipped with a BCI-integrated V2V communication system, other vehicles can receive the information and related prompts in time to take measures to avoid collisions. This cooperative driving information obtained through BCI technology can optimize traffic flow, reduce congestion, and prevent traffic accidents.

The application of BCI technology in vehicle communication systems is not a distant prospect but a technology with clear theoretical foundations that has entered the substantive research stage. The construction of such ITS is not technically complex and only requires three steps.

(1) *Real-time vehicle monitoring*: Real-time monitoring of the driver's EEG data within the area through BCI technology to identify the driver's emergency responses.
(2) *Information transmission*: When the driver issues an emergency command (such as emergency braking or evasive maneuvers), the BCI system transmits the information to nearby vehicles within the area via V2V communication.
(3) *Timely adjustment*: After receiving the information, nearby vehicles within the area promptly adjust their speed and driving paths to ensure coordinated and safe traffic throughout the area. Through this intelligent driving arrangement, vehicles can maintain consistent speeds and distances, reducing traffic accidents caused by speed inconsistencies and improving overall traffic efficiency.

In fact, the integration of BCI technology and V2V communication is not complex overall. Regardless of the application scenario, its technical logic follows the following three steps, as shown in Figure 8.1.

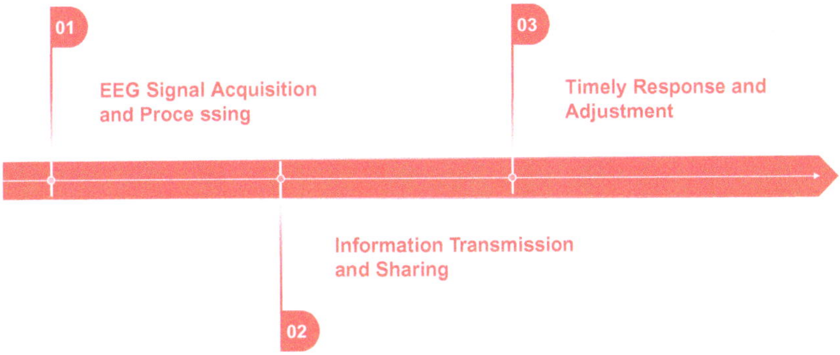


**Figure 8.1.** Technical logic steps for the integration of brain–computer interface technology and V2V communication.

(1) *EEG Collection and Processing*: Real-time collection of the driver's EEG signals through EEG devices, using signal processing algorithms to analyze the driver's state and intentions.
(2) *Information Transmission and Sharing*: Through the V2V communication protocol, the driver's status information and emergency commands are transmitted to other vehicles to achieve information sharing.
(3) *Timely Response and Adjustment*: Vehicles receiving information respond and adjust through the autonomous driving system to ensure overall traffic safety and smoothness.

In this process, the BCI system requires efficient signal processing capabilities and rapid information transmission to ensure that the driver's status information can be transmitted in real time and responded to quickly. Meanwhile, the V2V communication system needs to have high reliability and stability to guarantee the accuracy and timeliness of information transmission. These two key points are also the current technical challenges that need to be overcome.

However, the integration of BCI technology and V2V communication has become an important direction in the development of China's transportation technology. This integration will also become a crucial component of ITS, providing strong technical support for achieving safer, more efficient, and smarter traffic management, bringing more convenience and safety to people's travel.

## 8.5 Brain–Computer Interface and Traffic Planning

Traffic planning is the cornerstone of urban development, playing a vital role in managing urban traffic flow, designing road layouts, and optimizing public transportation systems. Traffic planning not only determines the efficiency of urban transportation but also directly affects residents' quality of life. With the acceleration of urbanization, traffic problems have become increasingly prominent, making reasonable traffic planning especially important.

In existing traffic planning methods, traditional traffic models and data analysis approaches dominate. These methods rely on historical data and long-term forecasts, typically using traffic demand prediction, traffic flow models, and simulation techniques. In the process of constructing China's integrated three-dimensional transportation network, traffic planning also requires more technological innovation to meet the overall demands of intelligent traffic development.

Here, it is important to note that traffic planning is not merely a data analysis process but a dynamic system requiring real-time adjustment and optimization. Changes in actual conditions, such as emergencies, weather changes, and traffic accidents, all demand that traffic planning possess high flexibility and real-time responsiveness. Currently, the development of transportation technology in China is also striving to enhance dynamic traffic planning capabilities, with BCI technology being a key force employed by China's transportation technology research departments. By collecting and analyzing EEG data from drivers and pedestrians, it provides new solutions for the real-time and dynamic adjustment of traffic planning.

In fact, the integration of BCI technology and traffic planning is not only a current research focus in China's transportation technology but also a project pursued by many developed countries worldwide, all designing the fusion of BCI technology with traffic planning.

The United States Defense Advanced Research Projects Agency (DARPA) is currently conducting a "Neural Engineering System Design" (NESD) project aimed at developing a high-resolution neural interface system. This system can achieve precise communication between the human brain and the digital world, with the core goal of enabling parallel interaction with over one million neurons through a high-density electrode array. The project is expected to cost $60 million and aims to comprehensively enhance the level of BCI technology in the United States, thereby providing new technical support for future traffic planning.

The core of the NESD project lies in improving the capacity and accuracy of neural interfaces, achieving efficient communication between the brain and machines through more advanced hardware and software. In practical applications, the NESD project demonstrates the potential of brain–machine interface technology in traffic management. For example, by monitoring the driver's EEG signals, the traffic management system can identify the driver's fatigue state and dynamically adjust traffic signals to prevent accidents. This research project can enhance the responsiveness and flexibility of the U.S. transportation system, providing technical support for the construction of ITS.

According to information released by the U.S. DARPA, the current challenges of the NESD project lie in solving the biocompatibility issues of high-density electrode arrays and ensuring the real-time and accuracy of data processing. In addition, data privacy and security are also important considerations. However, overall, the research and development progress of this project is relatively ideal.

There is also the European Union's "BrainCom" project. The "BrainCom" project is part of the "H2020 Program." The full name of the "H2020 Program" is the "Horizon 2020" Research and Innovation Framework Program. Its predecessor was the "EU Framework Program for Research," which started in 1984. It is a mid-term major research program jointly participated by EU member states and one of the largest official comprehensive science and technology research and development programs in the world, mainly focusing on cutting-edge international and competitive scientific and technological challenges.

The "BrainCom" project focuses on developing high-density EEG signal acquisition and processing technologies to enhance the performance of brain–machine interface systems. The research of this project not only provides new solutions for the medical field but also offers technical support for traffic planning and management.

According to relevant academic reports from Europe, the BrainCom project is committed to improving the accuracy and real-time performance of brain–machine interface systems through high-density signal acquisition devices and advanced signal processing algorithms. For example, the devices developed by this project can densely collect EEG signals and, through data fusion technology, combine these signals with other traffic data to provide comprehensive information support for traffic planning. In terms of application in traffic planning, the BrainCom project demonstrates its technological potential. Through high-density signal acquisition technology, it can monitor and analyze drivers' EEG in real time, dynamically adjust traffic management, reduce traffic congestion, and improve traffic flow efficiency. This technology can also be used to monitor pedestrian status, providing a safer traffic environment. The scientific research results of this project have fully highlighted the value of brain–machine interface technology in the field of traffic planning and management.

Additionally, the brain–machine interface project at Japan's RIKEN Brain Science Institute has also been mentioned in related scientific reports. The project develops high-precision EEG signal processing and real-time data analysis technologies, which can effectively improve the efficiency and safety of traffic management, thereby achieving smarter traffic planning and management outcomes.

From the above research projects, we can see the enormous potential of brain–machine interface technology in traffic planning. These projects demonstrate the application strategies and technical focuses of brain–machine interface technology in real-time data acquisition and analysis, dynamic traffic management and optimization, personalized traffic services, and multi-modal data fusion, bringing multi-dimensional promotion to the intelligent management of traffic systems. Research and applications in this field across various countries all indicate that brain–machine interface technology will become an important technological force in traffic planning and management.

Combining the research and development achievements of brain–machine interface technology in traffic planning worldwide, as well as the enhancements and promotions by scholars globally regarding the application of brain–machine interface technology in traffic planning, these can be reflected in the following aspects, as shown in Figure 8.2.

1. **Traffic Demand Analysis:** Traffic demand analysis is the foundation of traffic planning, and traditional methods mainly rely on surveys and statistical data. However, BCI technology can more accurately understand the travel demands and behavior patterns of drivers and pedestrians by directly reading and analyzing their brainwaves. By monitoring drivers' travel intentions through BCI devices, travel time and route choices can be optimized.

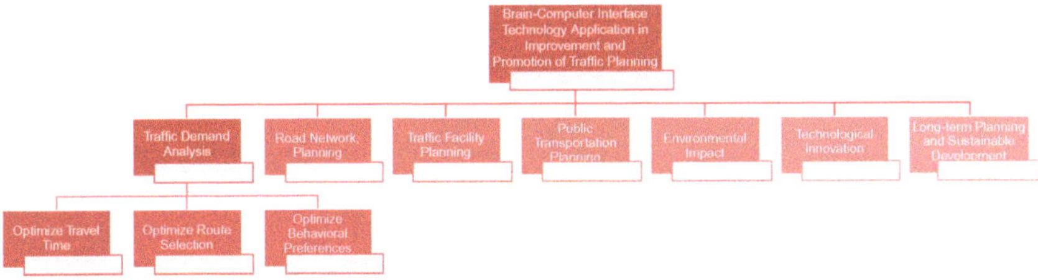


**Figure 8.2.** Enhancement and promotion of brain–machine interface technology applied to traffic planning.

At the same time, BCI technology also helps researchers better understand the travel preferences of different groups, providing more reliable data support for traffic demand forecasting.

**2. Road Network Planning:** In road network construction planning, the application of BCI technology can enhance the scientific and rational aspects of planning. Through brainwave data, planners can more intuitively understand drivers' psychological and physiological responses under different road sections and traffic conditions. For example, when designing highways and urban main roads, BCI technology can assist designers and constructors in collecting drivers' feedback on lane width and curvature to determine more comfortable and safer lane design schemes. At the same time, BCI technology can be used to simulate and evaluate the actual effects of different road network plans, thereby determining the optimal design outcome.

**3. Transportation Facility Planning:** Transportation facility planning includes the layout and design of facilities such as parking lots and service areas. The application of BCI technology in this field mainly focuses on optimizing user experience and improving facility utilization efficiency. By monitoring drivers' brainwave activity, BCI technology can help determine the optimal rest distances and locations, provide reasonable rest suggestions, and prevent driving fatigue. In addition, in parking lot planning, BCI technology can help analyze drivers' stress and anxiety when searching for parking spaces, optimize the design and guidance systems of parking lots, and improve parking efficiency.

**4. Public Transportation Planning:** Public transportation is an important means to solve urban traffic problems, and the application of BCI technology in public transportation planning mainly focuses on enhancing passenger experience and operational efficiency. By monitoring passengers' brainwave activity, BCI technology can help identify passengers' comfort and satisfaction levels with different transportation modes and services, thereby optimizing the design and operational strategies of buses, subways, light rails, and other transit vehicles.

**5. Environmental Impact:** With the increasing awareness of environmental protection, environmental impact assessment in transportation planning has become increasingly important. The application of BCI technology in this area mainly focuses on monitoring and mitigating the effects of traffic pollution on people. By monitoring pedestrians' brainwave activity in different environments, BCI technology can help assess the impact of traffic noise and air pollution on residents' health and psychology, and formulate more effective emission reduction and noise control measures. For example, when planning highways and railway lines, BCI technology can assist designers in collecting relevant information to determine the most appropriate layout of sound barriers and green belts to reduce noise disturbance to residents. In addition, BCI technology can also be used to evaluate the environmental performance of different transportation modes, promoting the development and application of low-carbon vehicles.

**6. Technological Innovation:** BCI technology itself is a cutting-edge innovation. Its application in transportation planning not only enhances the intelligence level of transportation systems but also gives rise to many new application scenarios and service models. For example, the combination of BCI technology and V2V communication technology can achieve real-time communication and cooperative driving between vehicles, improving traffic flow and safety.

**7. Long-term Planning and Sustainable Development:** In terms of long-term planning and sustainable development, the application of BCI technology helps build smarter and more sustainable transportation systems. By continuously monitoring and analyzing the brainwave data of traffic participants, BCI technology can provide real-time traffic flow and demand forecasts, offering a scientific basis for transportation planning and management.

Overall, the application prospects of BCI technology in transportation planning are broad. It not only enhances the scientific and precise nature of transportation planning but also brings new possibilities for the intelligence of transportation systems and user experience. We can see that as the application of BCI technology deepens in the transportation field, it will gradually become an important tool for transportation planning and a key support for building more efficient, safe, and sustainable transportation systems.

## 8.6 Brain–Computer Interface and Travel for People with Disabilities

Travel is a fundamental right for everyone, including people with disabilities. Being able to travel independently means they can freely participate in various social activities such as work, study, shopping, and entertainment. This not only helps improve their quality of life but also allows them to feel equality and respect. People with disabilities who lack travel ability face more barriers in society, not only physically restricted but also feeling isolated and helpless psychologically.

Travel ability directly determines the social integration of people with disabilities. Those who can travel conveniently are more likely to find jobs, receive education, access medical services, and participate in community activities. This has a positive impact on their mental health and sense of social identity, helping to enhance their self-worth and happiness.

Although solving travel issues is crucial for people with disabilities and our country has made great efforts in this regard—such as establishing dedicated travel channels for people with disabilities, providing related services at public transportation hubs, and setting priority seats for people with disabilities on vehicles—many challenges still exist in the actual travel of many people with disabilities. Addressing these problems remains a task for the development of our transportation system.

First, physical barriers are a major issue. Some public transportation and infrastructure are still not friendly to people with disabilities, lacking barrier-free access and equipment. For example, many buses and subway stations are not equipped with wheelchair ramps, elevators, or automatic doors. People with disabilities often need to rely on others for assistance during travel, which greatly limits their independence and freedom. In some cities, even where barrier-free facilities exist, problems with maintenance and management prevent people with disabilities from using them smoothly.

Second, technical limitations are also an important factor. Existing assistive technologies have limitations in functionality and ease of use, failing to fully meet the travel needs of people with disabilities. Many assistive devices such as wheelchairs and walkers, while helpful to some extent, still have significant shortcomings in flexibility, usability, and adaptability. For example, operating a wheelchair depends on hand strength and dexterity, which is a huge challenge for some people with disabilities. Although electric wheelchairs offer greater flexibility, they still face many restrictions in complex urban environments, such as narrow roads, uneven surfaces, and uneven curbs.

Finally, economic burden is also a major obstacle faced by people with disabilities. The high cost of assistive devices and services becomes an economic barrier to their travel. Many high-tech assistive devices are expensive and unaffordable for ordinary families. Even in developed countries, people with disabilities often face difficulties in paying for these devices and services, further limiting their travel ability and quality of life. In addition, the maintenance and repair costs of assistive devices often exceed what people with disabilities and their families can afford, forcing many to rely on cheap but functionally limited equipment.

In recent years, with the development of transportation technology in China, the travel issues of people with disabilities have been greatly improved. Moreover, in terms of transportation facilities and vehicles, the transportation authorities in China have also carried out humanized upgrades and improvements, which have significantly enhanced the safety and independence of people with disabilities during travel, as well as improved their quality of life and social participation. However, with the integration of BCI technology and transportation technology, the travel problems of people with disabilities will be further improved in multiple aspects.

### 1. Improvement of Mobility Tools Operation for People with Disabilities

Through BCI technology, people with disabilities can directly control devices using brainwave signals without relying on physical movements, overcoming physical barriers. This significantly enhances the possibility and safety of independent mobility, especially for those who have lost basic operational abilities.

For example, an intelligent wheelchair controlled via BCI allows people with disabilities who cannot use hand strength to operate the wheelchair solely through their mental intentions. This enables them to perform daily activities more independently, such as shopping, working, and socializing. The emergence of intelligent wheelchairs not only improves their mobility but also enhances their independence and self-esteem. Currently, multiple research applications of BCI technology in the medical field have been carried out in China. Related studies show that people with disabilities controlling wheelchairs using BCI technology experience unprecedented freedom and autonomy during travel. They can freely control the direction and speed of the wheelchair, greatly improving travel efficiency and safety.

Shanghai Shuli Intelligent Technology Co., Ltd. (referred to as Shuli Innovation Group) is a company established in 2015, primarily engaged in research and experimental development. In recent years, the company has focused on the development of non-invasive BCI technology. This technology has achieved significant results in the diagnosis and treatment of motor disorders, neurodevelopmental disorders, and mental illnesses.

In May 2023, a 38-year-old "programmer" was admitted to the hospital due to a sudden cerebral hemorrhage with a bleeding volume of 40 milliliters, resulting in unclear speech, right-sided hemiplegia, and significant decline in comprehension. Due to the severity of the condition, many hospitals considered his chances of recovery slim, and his family was nearly hopeless.

In June 2023, the patient began using Shuli Innovation's BCI diagnostic and treatment system. The system extracted the patient's motor intention signals through an EEG cap and transmitted these signals to external devices, using an exoskeleton to drive limb movement. Feedback signals were transmitted back to the brain via peripheral nerves, forming a closed-loop neural conduction system. Through repeated training, the brain gradually formed new control areas to replace the damaged central nervous control regions.

During the 7-week treatment period, the patient underwent 42 training sessions. His limb functions gradually recovered, and speech impairments significantly improved. Upon discharge, his muscle strength was close to normal, fine motor skills of the right hand markedly improved, speech function returned to normal, and he was ultimately able to return to work.

The application of Shuli Innovation Group's non-invasive BCI technology in the mobility of people with disabilities has significantly improved their quality of life, enhancing independence and social participation. This technology demonstrates the great potential of BCIs in the field of mobility for people with disabilities.

**2. Intelligent Navigation for People with Disabilities**

Intelligent navigation systems combined with BCI technology can provide precise travel support for visually impaired individuals. Through brainwave signals, the system can analyze the user's intentions and provide real-time navigation instructions, helping them avoid obstacles and dangerous areas.

The real-time feedback function of the intelligent navigation system enables visually impaired individuals to travel freely in complex urban environments, enjoying a more independent and autonomous life. For example, the intelligent navigation system can guide visually impaired users safely across streets, help them find bus stops, and avoid obstacles through voice prompts or vibration feedback, allowing them to travel alone with greater confidence.

Real-time feedback and adjustment are major advantages of BCI systems. BCI systems can monitor the user's status in real time and provide feedback. For example, when user fatigue is detected, the system can suggest rest or automatically adjust the device's operating mode to ensure travel safety. At the same time, the real-time feedback mechanism allows users to quickly adjust their operations, improving travel efficiency and safety. This intelligent real-time adjustment function enables people with disabilities to travel more securely, reducing the likelihood of accidents. Research has found that through real-time monitoring and feedback, BCI systems can effectively prevent fatigued driving and attention lapses, enhancing the travel safety of people with disabilities.

**3. Providing Psychological Support for People with Disabilities in Travel**

Psychological support is a potential benefit of BCI technology. For people with disabilities who have long relied on others for assistance, the independence brought by BCI technology is not only physical but also psychological support. By autonomously controlling travel devices, people with disabilities can regain a sense of control over their lives, enhancing their self-worth and social participation. The application of BCI technology brings more hope and confidence to people with disabilities, enabling them to face life's challenges positively.

**4. Improving the Quality of Travel for People with Disabilities**

With technological development, future EEG devices are expected to become more compact, portable, and efficient, improving the quality of brainwave data acquisition and user experience. Signal processing algorithms will become more advanced, capable of interpreting brainwave signals more accurately and rapidly. These technological innovations will further enhance the performance of BCI systems, enabling their application in more scenarios. Higher-precision EEG devices and more advanced signal processing algorithms will improve the accuracy of brainwave data acquisition and analysis, allowing BCI systems to better serve the travel needs of people with disabilities while improving their quality of life.

In summary, BCI technology holds great potential in improving travel for people with disabilities. The future application of this technology will provide more convenient and safer travel solutions for people with disabilities, promising to bring benefits to more individuals and further promote the development of ITS, achieving the goal of easy and normal travel for people with disabilities.

# Part III

# Application Layer

# Chapter 9

# The Role of Brain–Computer Interfaces in Traffic Safety

## 9.1 Brain–Computer Interfaces and Driver Fatigue Detection

Brain–computer interface (BCI), as an emerging technological approach, is considered an important means for future interaction between artificial intelligence and humans. In recent years, with the rapid development of neuroscience, biomedical engineering, and other fields, the application of BCI technology in driver fatigue detection has increasingly attracted attention.

**1. Brain–Computer Interface Technology Principles**

BCI technology is a technique that captures brain signals through non-invasive methods and converts them into recognizable commands or information. Its fundamental principle is to combine signals generated by brain activity (such as electroencephalogram, EEG) with specific algorithms to identify and interpret the user's intentions. In driver fatigue detection, BCI technology can monitor the driver's EEG activity in real time to determine their fatigue state.

**2. Application Methods of Brain–Computer Interface in Driver Fatigue Detection**

(1) *EEG Signal Acquisition*: Acquire the driver's EEG signals through a non-invasive EEG. Compared with traditional fatigue detection methods,

EEG signals have advantages such as real-time capability, non-contact nature, and high sensitivity.

(2) *Signal Preprocessing*: Perform preprocessing operations on the collected EEG signals, such as denoising, amplification, and filtering, to improve signal quality and facilitate subsequent analysis.

(3) *Feature Extraction*: Extract meaningful feature parameters from the preprocessed EEG signals, such as frequency domain features and time domain features. These feature parameters can be used to characterize the driver's fatigue state.

(4) *Fatigue State Recognition*: Using machine learning, deep learning, and other methods to classify the extracted feature parameters, achieving the recognition of driver fatigue states.

To improve traffic safety, a company developed a driver fatigue detection system based on BCI technology to help drivers timely detect and respond to fatigue driving conditions. The system monitors the driver's brain activity in real time through a BCI device embedded in the driver's helmet. It uses deep learning algorithms to analyze the driver's EEG signals and identify characteristic patterns of fatigue states.

When the system detects signs of driver fatigue, it alerts the driver to rest promptly via the in-vehicle display or audio prompts. At the same time, the system automatically takes measures such as adjusting seat vibrations and automatic deceleration to ensure the driver's safety.

After field testing, the driver fatigue detection system performed excellently in recognizing fatigue driving states, successfully preventing multiple potential traffic accidents. Drivers also reported that the system's accuracy and timeliness of alerts made them feel more secure and confident while driving.

**3. Advantages of Brain–Computer Interface Technology in Driver Fatigue Detection**

(1) *Real-time Capability*: BCI technology can monitor the driver's EEG activity in real time, thereby judging the fatigue state instantly and providing timely warnings to the driver.

(2) *Non-contact Nature*: Compared to traditional fatigue detection methods (such as heart rate and blood pressure), BCI technology uses non-contact EEG devices to collect signals, reducing interference to the driver.

(3) *High Sensitivity*: BCI technology is highly sensitive to subtle electroencephalographic changes in drivers, which helps more accurately assess the fatigue state of the driver.

(4) *Personalized Customization*: BCI technology can tailor fatigue monitoring strategies according to the individual differences of each driver. In this way, drivers can choose the fatigue monitoring method that best suits their own needs and characteristics, enhancing the specificity and effectiveness of the monitoring.

(5) *Multifunctionality*: BCI technology can not only monitor the driver's fatigue status in real time but also track other physiological indicators such as emotions and attention. This will help comprehensively assess the driver's condition and provide more reference data for preventing traffic accidents.

(6) *Strong Adaptability*: BCI technology has strong adaptability and can operate normally in various driving environments, such as high temperature, low temperature, and humid harsh conditions. This enables the technology to provide stable fatigue monitoring services for drivers across different regions and seasonal contexts.

(7) *Data Confidentiality*: BCI technology uses encryption algorithms to ensure the security of data when collecting and processing drivers' EEG data. This helps protect drivers' privacy and prevents risks associated with data leakage.

(8) *Trend Toward Intelligent Development*: With the continuous development of artificial intelligence technology, BCI technology will achieve more intelligent and automated fatigue monitoring. In the future, the monitoring of drivers' fatigue status may be entirely handled by BCI systems, thereby reducing errors caused by human intervention.

BCI technology has significant advantages in monitoring driver fatigue. However, we should also recognize that it is still in the development stage, and some technical challenges such as signal interference and data processing speed need to be overcome in practical applications. It is believed that in the near future, BCI technology will play a greater role in the development of traffic safety in China.

## 9.2 Brain–Computer Interface and Driver Distraction Detection

Brain–computer interfaces (BCIs) also play a significant role in detecting driver distraction.

**1. Real-time Monitoring of Driver Attention:** BCI technology can monitor the driver's attention status in real time by collecting and analyzing brain signals to determine whether the driver is distracted. When the driver's attention is not focused, the system issues an alert to remind the driver to concentrate, thereby reducing the occurrence of traffic accidents.

**2. Personalized Driving Assistance System:** By analyzing the characteristics of the driver's EEG signals, a personalized driving assistance system can be tailored for each driver. When the driver is distracted, the system can automatically adjust the vehicle's speed, driving trajectory, and other parameters to ensure driving safety.

**3. Fatigue Driving Warning:** Long periods of driving can easily lead to driver fatigue, which in turn causes distraction. BCI technology can monitor the driver's fatigue level in real time and issue warnings when fatigue is detected, reminding the driver to rest in time to avoid traffic accidents.

**4. Improving Driver Training Effectiveness:** BCI technology can be applied in the field of driver training by analyzing the trainee's EEG signals to understand their attention, emotions, and other states during driving, providing targeted teaching suggestions for instructors and enhancing the training effectiveness.

The intelligent human–machine system team led by Professor Bi Luzheng from the School of Mechanical and Vehicle Engineering at Beijing Institute of Technology, with Assistant Professor Luo Longxi and PhD student Ju Jiawei as the main leaders, developed an intelligent driving assistance system called IDAS. This system combines EEG and electromyogram (EMG) signals to accurately classify the driver's braking and normal driving intentions. When an emergency is detected, the system can indirectly influence vehicle control or directly control the vehicle after confirming the emergency, significantly enhancing driving safety.

IDAS primarily receives information related to the vehicle and environment, driver behavior, and biological signals. Among these, the collection of biological signals is mainly achieved by monitoring the driver's foot, limb movements, and neural activities. Notably, they employed a hybrid brain–computer interface (hBCI) technology that combines EEG and other signals (such as EMG and electrocardiogram (ECG) signals) to improve the stability and performance of the BCI.

In the experiment, the team invited 13 subjects aged between 24 and 30 to participate. By collecting EEG, EMG, and vehicle information in simulated driving scenarios, they successfully used the hBCI model to detect the driver's emergency braking intention, as shown in Figure 9.1.

Additionally, the team compared several synchronous and sequential hBCI models, using different feature extraction and classification strategies to perform multi-class classification of three driving intentions (normal driving, soft braking, and hard braking). The results showed that the hBCI-SE1 classification algorithm based on spectral features and the one-vs-one classification strategy achieved the highest classification accuracy, with the system's average accuracy reaching 96.37%.

This study provides a reference for human-centered IDAS, helping improve driving safety and comfort. However, the project currently has some limitations, such as stimulus factors, subject variability, and data acquisition equipment. The team plans to further address these issues and

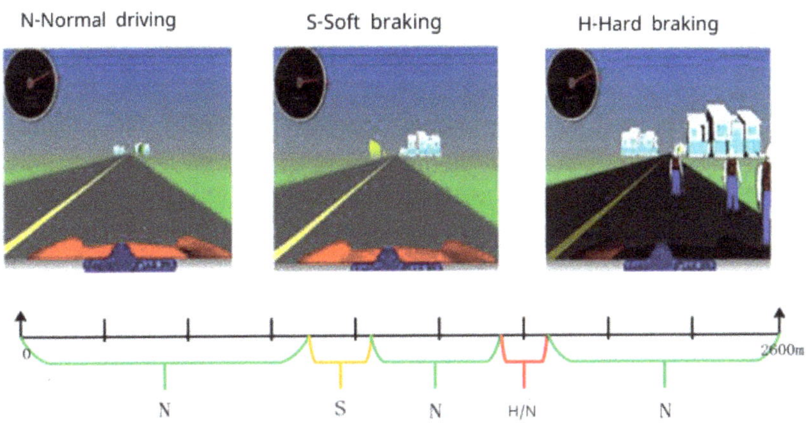


**Figure 9.1.** The hBCI model detected the driver's emergency braking intention.

explore more effective feature and strategy fusion to enhance system performance.

It can be seen that BCI technology has great potential in driver distraction detection, but it still faces some technical challenges, such as signal acquisition and processing, data analysis, and system real-time performance. Future research needs to focus on brain signal processing methods to improve detection accuracy and real-time capability.

Moreover, the application of BCI technology in driver distraction detection involves legal, regulatory, and ethical issues. How to protect driver privacy while ensuring driving safety is a problem that BCI technology must overcome in practical applications.

BCI technology has not yet been widely applied in driver distraction detection, partly due to high costs and partly because of limited acceptance of new technology by drivers. Reducing costs, improving market competitiveness, and expanding the application of BCI technology in driver distraction detection are key to future development.

With the continuous development of BCI technology, its application in driver distraction detection will become more widespread. In the future, the integration of BCI technology with artificial intelligence, big data, and other technologies will provide stronger guarantees for road traffic safety. At the same time, as autonomous driving technology matures, BCI technology will also play an important role in the field of unmanned driving.

## 9.3 Brain–Computer Interface and Driver Emotion Recognition

BCI technology can recognize the emotional states of drivers, such as anger and anxiety, which may cause driver distraction. The system can help drivers regulate their emotions and maintain a calm mindset through functions such as voice prompts and seat massage, thereby improving driving safety.

Driver emotion recognition and regulation technology has significant application value in the field of intelligent transportation. Through this technology, we can monitor the driver's emotional state in real time and take timely measures to reduce the occurrence of traffic accidents.

First, driver emotion recognition technology is mainly realized through BCI technology. By monitoring the driver's EEG signals, the system can determine the driver's current emotional state, such as anger or anxiety. These emotions are often important factors leading to traffic accidents; therefore, timely recognition of the driver's emotional state is crucial.

Once the system detects abnormal driver emotions, it will take a series of intervention measures to help the driver regulate emotions and maintain a calm mindset. These interventions include issuing voice prompts to remind the driver to adjust their emotions; using functions such as seat massage to relieve the driver's tension; and even playing relaxing music inside the vehicle to help the driver relax.

Moreover, driver emotion recognition and regulation technology can be integrated with other IDAS, such as autonomous driving and lane departure warning. These systems can complement each other to jointly improve driving safety. For example, when the system detects abnormal driver emotions, it can automatically adjust the vehicle speed, maintain a safe distance, and avoid traffic accidents.

As driver emotion recognition and regulation technology gradually develops, automobile manufacturers and technology companies are also continuously increasing their research and development efforts on IDAS systems. These systems not only improve driving safety but also help reduce driver fatigue and stress to some extent. For instance, the lane departure warning system can alert the driver when unintentionally drifting out of the lane to prevent accidents; the autonomous driving system can complete driving tasks automatically in complex road conditions, reducing the driver's driving difficulty.

However, relying solely on technological means is not enough to ensure road traffic safety. Driver emotion regulation and road safety education are equally important. Therefore, it is also crucial to strengthen the promotion of driver emotion regulation and road safety education. Using various media channels to raise public awareness about the importance of driver emotion regulation, reminding drivers to pay attention to their emotional changes, and encouraging them to take appropriate regulation measures when experiencing emotional fluctuations are very important approaches.

In addition, driving schools and driver training institutions are gradually incorporating emotion regulation and road safety knowledge into their teaching content. While learning driving skills, trainees also learn how to regulate emotions and respond to emergencies. In this way, trainees can establish good emotional regulation and road safety awareness before obtaining their driver's license.

With the development of artificial intelligence and big data technologies, driver emotion recognition and regulation technology will become increasingly mature. In the future, this technology is expected to become a standard feature in vehicles, providing drivers with a safer and more comfortable driving experience.

Driver emotion recognition and regulation technology has broad application prospects in the field of intelligent transportation. By monitoring the driver's emotional state in real time and taking timely intervention measures, we can effectively reduce the occurrence of traffic accidents and ensure road traffic safety. With the continuous advancement of related technologies, it is believed that this concept will gradually become a reality, making our travel better.

# Chapter 10

# The Impact of Brain–Computer Interfaces on Driver Attention

## 10.1 The Impact of Brain–Computer Interfaces on Driver Visual Attention

Traffic safety has always been a focus of attention across all sectors of society. With the continuous increase in vehicle ownership and the complexity of road networks, the frequency of traffic accidents has also risen. According to statistics, in 2023, there were 1.75 million road traffic accidents nationwide, an 8% increase compared to 2022. Among these, the number of deaths reached 500,000, up by 4%. The number of injuries reached 60,000, up by 5%. The national road traffic accident death rate per 10,000 vehicles was 3.6, basically unchanged from 2022. These data all reflect the severity of the current road traffic safety situation.

Globally, road traffic accidents are also a major problem. According to the World Health Organization (WHO), approximately 1.35 million people die annually worldwide due to road traffic accidents, and 20 to 50 million people are injured or disabled as a result. Most traffic accidents are caused by factors such as driver inattention, fatigue driving, and drunk driving. Among these, driver distraction and unstable mental and emotional states are important causes of traffic accidents. Therefore, improving drivers' attention levels, especially visual attention, is crucial for reducing traffic accidents.[1]

[1] The World Health Organization (WHO), commonly referred to in Chinese as the World Health Organization, is a specialized agency of the United Nations. Headquartered in

Visual attention refers to our selective processing ability of visual information within a specific period. During driving, drivers need to perceive road conditions, traffic signals, pedestrians, and other vehicles through vision, so as to make timely and accurate driving decisions. The importance of visual attention in the driving process is self-evident.

The level of visual attention directly affects the driver's ability to perceive road conditions and reaction speed. When attention is focused, drivers can more quickly identify potential dangers on the road and take corresponding avoidance measures; when attention is distracted, drivers may miss important visual information, thereby increasing the risk of accidents.

In traditional driving methods, drivers mainly rely on visual attention to operate vehicles. Drivers observe road conditions with their eyes and transmit this visual information to the brain for processing and analysis. After receiving the visual information, the brain makes corresponding decisions based on the current driving environment and experience, such as accelerating, braking, and steering.

However, in modern society, factors causing driver visual attention distraction are increasing. Mobile phone use is a significant example.

On April 6, 2024, the American Popular Science website published an article titled *Do You Use Your Phone While Driving?* The article mentioned that fatal crashes caused by driver inattention have not significantly decreased over time. In 2017, fatal crashes caused by inattention accounted for 14%. Although various related policies have been introduced in the United States since then, the situation has not improved, and the proportion remained 13% in 2021. Moreover, experts believe these figures are based on police-reported crashes, and the actual situation is likely far worse.

Chinese traffic experts have also conducted research on this issue. According to expert estimates, the accident rate caused by mobile phone use during driving is 23 times that of normal driving. That is, when drivers pay attention to phone information, answer calls, or operate their phones

---

Geneva, Switzerland, it is the world's largest intergovernmental health organization, with membership limited to sovereign states. The mission of the WHO is to ensure that all people worldwide attain the highest possible level of health. Its primary functions include promoting the prevention and control of epidemics and endemic diseases, enhancing public health education and training in disease treatment and related matters, and advancing the establishment of international standards for biological products.

during driving, the probability of a traffic accident is 23 times higher than when not paying attention to the phone. The study also includes a set of data that clearly explains the severity of this situation.

Taking a vehicle speed of 60 km per hour as an example, it takes at least 3 seconds to pay attention to phone information while driving, which means the car will travel 50 m in a blind driving state. At 60 km per hour, the braking distance of the car is at least 20 m, which means watching the phone for 3 seconds while driving will greatly increase the accident risk over a distance of 70 m.

In reality, whether making phone calls, sending text messages, or using navigation applications, the driver's attention is diverted from the road, significantly increasing the risk of accidents. Although navigation systems improve driving convenience to some extent, they can also become a source of distraction for drivers. When drivers frequently check navigation devices, their visual attention is scattered and cannot focus on road conditions.

Interaction with passengers inside the vehicle is also an important factor. Studies show that when drivers communicate with passengers, especially during intense discussions or arguments, their attention easily shifts away from the driving task, leading to delayed reactions. In addition, in-car entertainment systems, such as onboard televisions and audio systems, also interfere with the driver's visual attention.

Furthermore, psychological and physiological factors also cause drivers' visual attention to be distracted. Psychological factors such as stress, anxiety, and emotional fluctuations affect the driver's attention level. When drivers are in a highly tense or emotionally unstable state, their ability to perceive external information significantly decreases, and reaction times lengthen, thereby increasing the risk of traffic accidents. Physiological factors such as fatigue and lack of sleep also severely impact the driver's attention level. Long-term driving easily leads to driver fatigue, difficulty concentrating, and even drowsiness.

To improve this situation, researchers in China's transportation field have begun applying BIC technology to enhance driver attention, with visual attention improvement being the primary research focus.

Currently, there are numerous research projects in China applying brain–computer interface (BCI) technology to the driving field, and enhancing driver attention, especially visual attention, is a key focus of these projects.

**Case 1: Integration of Brain–Computer Interface and Intelligent Vehicles at National University of Defense Technology**
On December 24, 2021, a patent titled "A Vehicle Control Method and Device Based on Brain–Computer Interaction" from the National University of Defense Technology was authorized. This patent combines BCI technology with intelligent vehicles, enabling operations such as overtaking, following, left turn, right turn, and roundabout navigation in complex driving environments. Through this system, drivers can maintain a high level of visual attention while controlling the vehicle more efficiently and safely.

According to the patent description, the system mainly includes the following four modules:

(1) *Information perception*: During driving, the vehicle uses its own sensors to perceive external environmental information and the vehicle's own status information in real time. This information includes surrounding obstacles, road conditions, vehicle speed, and vehicle position.
(2) *Decision demand judgment*: The system determines all driving scenarios that match the current environment based on real-time perception of external environmental information and the vehicle's own status information and generates optional driving decision outputs for the driver to choose from. For example, when approaching an intersection, the system can provide options such as going straight, turning left, or turning right.
(3) *Decision command extraction*: The system collects the driver's electroencephalogram (EEG) signals in real time and extracts the driver's decision-making commands from them. By analyzing these EEG signals, the system can accurately determine the driver's intentions and convert them into specific driving commands.
(4) *Vehicle control driving*: Upon receiving the driver's decision-making commands, the system calls the corresponding target model from multiple pre-trained driving behavior models. These driving behavior models include steering control information of the driver under different driving scenarios. The system uses the target driving behavior model to control the vehicle's driving behavior. For example, during overtaking, the system adjusts the vehicle speed and steering angle according to the driver's intention to safely complete the overtaking maneuver.

In simulation tests, the system performed excellently. The test environment included complex urban roads and highways, with the simulated vehicle speed ranging from 25 km/h to 50 km/h. The system was able to effectively complete functions such as obstacle avoidance, overtaking, and meeting oncoming vehicles, significantly improving the driver's visual attention and decision-making efficiency in complex driving environments.

In one simulation test, the vehicle was driving on an urban road when an obstacle suddenly appeared ahead. The system quickly sensed this situation through sensors and provided the driver with optional driving decisions (such as emergency braking or detouring). The driver selected the detour option via EEG signals, and the system immediately executed the corresponding driving behavior, successfully avoiding the obstacle. This demonstrated the advantage of BCI technology in enhancing the driver's visual attention and reaction speed.

In this way, BCI technology not only improves the driver's concentration level but also reduces their operational burden. During driving, the system helps the driver process a large amount of information through sensors and decision algorithms, allowing them to focus on critical decisions. This human–machine collaborative driving mode not only enhances driving safety but also provides new ideas for the future development of autonomous driving technology.

**Case 2: Brain 360 Focus Training Mode**

In the field of brain–computer interfaces (BCIs), a Brain 360 Focus Training Mode based on BCI technology has also emerged. This training mode aims to help individuals improve their focus level, enhance attention control ability, and boost cognitive performance. By combining advanced BCI technology with scientifically validated paradigms, Brain 360 Focus Training offers users a professional and personalized training experience.

Brain 360 Focus Training leverages BCI technology to achieve direct interaction with the user's brain activity. The system places EEG sensors on the scalp to capture and interpret the brain's electrical signals in real time, converting them into commands to interact with computers or mobile devices. This innovative technology provides accuracy and real-time responsiveness for Brain 360 Focus Training, making the training process more refined and personalized.

Through interaction with the BCI system, users can understand their focus level in real time and receive timely feedback and guidance during training. The training paradigm typically includes the following key elements:

(1) *Focus tasks*: Users need to maintain focus and concentration on specific tasks, such as observation, memory, or problem-solving. The type and difficulty level of the tasks are adjusted according to individual needs and abilities.
(2) *Feedback mechanism*: BCI system monitors the user's EEG activity in real time and provides timely feedback to the user through visual or auditory means. This allows the user to intuitively understand their focus state and make adjustments and improvements based on the feedback.
(3) *Training program*: The Brain 360 Focus Training develops personalized training plans based on the user's starting point and goals. Through progressive training, users can gradually improve their concentration levels and continuously advance through ongoing challenges.

Nowadays, the Brain 360 Focus Training model has been widely applied in multiple fields, including education and professional training. In schools and educational institutions, it is used to enhance students' learning effectiveness and attention concentration. In professional training, it is employed to improve employees' work efficiency and focus, thereby enhancing work quality.

Through Brain 360 Focus Training, users' visual attention levels are significantly improved. This training method not only strengthens users' concentration but also enhances their performance in daily life and work. For example, students can maintain longer attention spans during study, improving memory and comprehension; professionals can focus more on tasks at work, increasing efficiency and quality.

From the above cases, we can see the great potential of BCI technology in enhancing drivers' visual attention and improving driving safety.

From a technical perspective, through technologies such as EEG, BCI can detect drivers' brain activity. Research shows that brain activity patterns vary with different attention levels. BCI technology can analyze these brain activity data to assess drivers' attention states in real time and issue alerts when attention is distracted. This directly promotes the improvement of drivers' attention.

BCI technology not only monitors drivers' attention states but also integrates with vehicle driver-assistance functions to provide support. For example, when the system detects a decline in driver attention and multiple reminders fail to improve it, the driver-assistance system can temporarily take over certain driving tasks, such as lane keeping and automatic braking, thereby reducing the driver's operational burden and allowing better concentration.

In addition to monitoring drivers' brain activity, many researchers are currently studying the use of BCI technology to monitor drivers' gaze status, aiming to develop a new visual tracking technology that enhances drivers' visual attention.

Visual tracking technology is a technique for accurately detecting, extracting, recognizing, and continuously tracking moving targets. This technology collects key motion parameters during the target's movement, providing a solid foundation for subsequent processing and analysis, thereby enabling an in-depth understanding of the target's behavior.

Gaze tracking technology combined with BCI technology can monitor drivers' eye movements through high-precision sensors and analyze their gaze direction and focus in real time to accurately understand drivers' visual attention. It then provides corresponding alerts and assistance to improve drivers' visual attention concentration. This technology consists of multiple technical modules; with the coordinated operation of these modules, drivers' visual attention can be precisely grasped, and the BCI-integrated assistance system can offer additional driving support. These modules mainly include the following:

(1) *Eye Movement Monitoring*: Real-time capture of the driver's eye movements through infrared cameras or eye trackers and recording of their gaze trajectories.
(2) *Pupil Tracking*: Using infrared light to illuminate the pupil, tracking eye movements by changes in the position of the reflected light.
(3) *Corneal Reflection*: Monitoring the reflection points on the corneal surface to calculate the gaze direction.
(4) *Data Processing*: Synchronously processing eye movement data with EEG signals to extract feature signals related to spatial attention.
(5) Signal Alignment.

Using Dynamic Time Warping (DTW) algorithm to align different signal sources, ensuring data synchronization.

(6) *Feature Fusion*: Combining eye movement and EEG features, comprehensive features are extracted through multi-modal learning algorithms.
(7) *Attention Assessment*: Using machine learning algorithms to analyze gaze trajectories and EEG data to evaluate the driver's spatial attention state.
(8) *Support Vector Machine (SVM)*: Using an SVM classifier to identify focused and distracted attention states.
(9) *Feedback Mechanism*: When the system detects that the driver's gaze deviates from important areas or attention is distracted, it reminds the driver to refocus through auditory, vibration, and visual prompts.

Gaze tracking technology can not only monitor the driver's gaze direction but also combine with the BCI feedback mechanism to provide multi-sensory prompts. When the driver's gaze deviates from important areas, the system can remind them to pay attention through sound, vibration, and visual cues. For example, if the driver fails to note a vehicle in the blind spot while changing lanes, the system can emit a warning sound and display a warning icon in the field of view to help the driver refocus.

In addition, gaze tracking technology can be combined with Augmented Reality (AR) technology to overlay virtual information in the driver's field of view, providing real-time spatial information and thus creating a visual enhancement system. For instance, when the driver is driving at high speed, the AR system can display dynamic information about the road ahead in the field of view, such as upcoming curves, traffic signs, and the positions of other vehicles, helping the driver maintain a high level of spatial attention. The implementation of this visual enhancement system typically involves the following modules:

(1) *Data Acquisition*: Collect environmental data through onboard cameras and sensors to generate real-time three-dimensional environmental models.
(2) *Multi-sensor Fusion*: Combine data from cameras, LiDAR, and ultrasonic sensors to improve the accuracy of environmental perception.
(3) *Virtual Information Generation*: Use computer vision and AR technologies to generate virtual information and overlay it onto 3D environmental models.

(4) *Deep Learning*: Utilize deep learning algorithms to recognize road signs, pedestrians, and other vehicles, generating corresponding virtual prompts.
(5) *Display Control*: Overlay virtual information onto the driver's field of view through in-vehicle display devices or generate prompt messages.
(6) *Projection Technology*: Use high-precision projection technology to display virtual information in the driver's field of view, ensuring clarity and visibility of the information.

Through this process, the visual enhancement system can provide rich spatial information without interfering with the driver's normal line of sight, helping the driver improve their perception of the surrounding environment.

From the perspective of the deep application of BCI technology, this technology can not only enhance the driver's visual attention but also improve the driver's operation methods by controlling certain vehicle functions through mental commands, thereby reducing the demand for visual attention. Under mental control, the driver can directly control the vehicle through brain intentions. This mode of mental control is safer than manual operation with hands and feet because people tend to have delayed or incorrect reactions under stress, whereas mental control can avoid these situations, providing faster responses and thus reducing traffic accidents.

It is believed that with continuous technological advancement and deeper application, BCI systems will become an important component of intelligent transportation, providing strong support for achieving safer and smarter driving experiences. In the future, we have every reason to believe that BCI technology will play an even more significant role in the transportation field, making greater contributions to reducing traffic accidents and improving road safety.

## 10.2 The Impact of Brain–Computer Interfaces on Drivers' Auditory Attention

During driving, most people usually focus their attention on vision, believing that driving safety mainly depends on the driving environment seen by the eyes. However, auditory attention is equally important as it can provide drivers with more information to help them better judge and respond to road conditions.

In previous English textbooks, there was a well-known little story about auditory attention. A little boy loved listening to the sound of train tracks while the train was running and could sense the train's operating status through the rhythm and tone of the sound. Once, while riding a train, he heard an unusual sound from the tracks. Thanks to his keen auditory attention, he immediately informed the conductor. After inspection, the conductor found that the train indeed had a problem and took timely measures, preventing a potential accident and saving the lives of all passengers on board.

In fact, auditory attention is very important in our daily driving. Such stories are not only found in books.

At the beginning of 2023, Shen Han, a toll collector from Hubei Transportation Investment Group Co., Ltd. (Hubei Jiaotou), saved a vehicle full of people by identifying a safety hazard through hearing, quickly gaining widespread attention online. On the evening of January 30, 2023, a small passenger car entered the entrance lane of the Enshi Toll Station operated by Hubei Jiaotou's Western Hubei branch. Shen Han vaguely heard an abnormal sound from the vehicle. Carefully deducing from the sound, he suspected a problem with the car's tire and promptly reminded the driver: "There might be a nail stuck in the tire; please pull over and check." After getting out to inspect, the driver indeed found a large screw embedded in the left front tire, posing a serious safety hazard. Thanks to Shen Han's reminder, the driver replaced the tire in time, avoiding risks during high-speed driving.

Hearing is the driver's second pair of eyes. It helps the driver perceive threats from blind spots, provides an additional layer of information, and assists the driver in making more accurate and timely decisions. The concentration of auditory attention not only affects the driver's reaction speed but also directly impacts driving safety.

During driving, sound signals such as sirens, honks, and navigation prompts are important alerts. Timely capturing and responding to these sound signals can help drivers better grasp driving dynamics and reduce accident risks. Therefore, improving drivers' auditory attention has significant practical importance.

In fact, vision and hearing are the two major sensory systems that drivers rely on during driving. Although vision provides the primary information about road conditions, hearing can complement the limitations of vision by offering drivers a 360-degree perception of the environment. For example, drivers can perceive emergency vehicles approaching from behind through hearing, which is especially important in areas not covered by vision.

Through hearing, drivers can promptly understand the dynamic changes in the surrounding environment and thus make more accurate responses.

At night or under adverse weather conditions, the effectiveness of vision is greatly reduced, while hearing can compensate for this deficiency. On foggy days, drivers find it difficult to see distant roads and vehicles visually, but they can prepare to avoid hazards in advance by hearing sirens or horns. In addition, hearing can help drivers identify road conditions; for example, by hearing the friction sound between tires and the road surface, they can judge whether the road is slippery and adjust their driving speed and manner accordingly.

Moreover, auditory information has the characteristics of immediacy and continuity. Compared to visual information, which may be limited by viewing angle and lighting, auditory information can be transmitted continuously and is not restricted by direction. This feature makes hearing particularly important in driving, especially when drivers need to multitask. Auditory information can provide additional important cues without interfering with visual tasks.

Among the relevant information that drivers can receive through hearing, alert sounds are especially important as a key means of warning and guidance. These alert sounds play an irreplaceable role during driving, effectively enhancing traffic safety and reducing accidents.

For example, in highway traffic, common alert sounds include in-vehicle warning sounds, road construction warning sounds, emergency vehicle sirens, and traffic signal sounds. These alert sounds provide critical information in different situations, helping drivers respond promptly.

Such as seatbelt reminders, door ajar warnings, and turn signal alert sounds. These sounds ensure that drivers maintain the basic safety status of the vehicle before and during driving; road construction zones typically use warning sounds to remind drivers to slow down, ensuring the safety of construction workers and drivers; emergency vehicles such as ambulances, fire trucks, and police cars use sirens to alert other vehicles to yield, providing a clear path for emergency rescue; pedestrian signal sounds at crosswalks help coordinate passage between pedestrians and drivers, especially when visibility is limited.

Furthermore, different traffic domains have various auditory information alerts, and drivers' attention to these sounds can effectively prevent traffic accidents.

For instance, in waterway traffic, common alert sounds include ship horns, port operation warning sounds, and severe weather warning sounds. Ships communicate and warn through horn signals during navigation, such as docking, departing, and yielding, ensuring navigation safety; port operation areas use warning sounds to alert ships and workers to safety hazards, avoiding collisions and accidents; traffic management broadcasts or alarms remind ships to be aware of severe weather conditions and take precautionary measures in advance.

In air traffic, common alert sounds include cabin warning sounds, air traffic control command tones, and internal crew communication sounds. For example, seatbelt reminder sounds and warning sounds during takeoff and landing ensure the safety of passengers and crew members; pilots receive air traffic control instructions via radio to ensure the aircraft flies safely according to commands; crew members communicate through internal communication sounds to ensure coordinated flight operations.

All of these require drivers to maintain sufficient auditory attention. However, many drivers nowadays lack auditory attention during driving for various reasons, thereby ignoring these critical alert sounds and causing traffic accidents.

On June 6, 2018, Mr. Ge was riding Metro Line 8 at People's Square Station in Shanghai. Because he did not hear the train door closing alert sound, his head was caught by the door when boarding. Hospital diagnosis confirmed that Mr. Ge suffered a head injury.

In 2023, a driver in China was listening to music while driving and did not hear the horn of an electric motorcycle approaching from the right rear side. When he was about to make a right turn, the electric motorcycle, speeding from the right rear, could not brake in time and collided with the right side of the car, causing a serious traffic accident. The driver, immersed in the music, ignored important external warning sounds, which ultimately led to the tragedy. This incident shows that maintaining sufficient auditory attention is a crucial foundation for safe driving.

In the application of BCI technology to enhance drivers' attention, the effect of improving auditory attention is also significant. The real-time monitoring and feedback mechanisms of BCI technology help drivers maintain a high level of auditory attention during long driving periods, thereby reducing the risk of accidents. Moreover, the application of BCI technology in this field is broader, greatly promoting autonomous driving, vehicle networking, and the intelligent development of vehicles themselves.

**1. Application in the field of autonomous driving**

The development of autonomous driving technology has greatly freed drivers' hands and eyes but has also led to a significant decline in driver attention. Especially with the continuous advancement of autonomous driving levels today, some cars have already achieved autonomous driving under specified conditions. For example, some Tesla models can activate the newly developed "Elon" mode at speeds below 60 km/h, enabling Level 3 autonomous driving without the need for the driver to keep their hands on the steering wheel. In such cases, the driver's attention significantly decreases.

BCI technology can provide appropriate auditory alerts in such situations. When an autonomous vehicle is about to enter a complex or congested road section, the system can proactively alert the driver through sound, allowing them to prepare to take over the vehicle, thereby improving driving safety.

**2. Application in vehicle networking interaction**

In intelligent transportation systems, BCI technology can be deeply integrated with vehicle networking. When an emergency occurs on the road, the system can detect emergency brain signals through monitoring the driver's emotional state and upload this information to traffic management authorities or the vehicle network system. At this time, the system can issue auditory alerts to surrounding vehicles, informing drivers about which lanes nearby have emergencies and require timely avoidance. For example, if a vehicle suddenly loses control on the highway, the system can immediately send emergency sound alerts to surrounding vehicles, prompting other drivers to avoid the area promptly and reduce the likelihood of accidents.

The combination of vehicle networking and BCI technology can achieve more intelligent and efficient traffic management, improving overall road safety and traffic flow efficiency. It is believed that with continuous technological progress, the application of BCI technology in vehicle networking will become more widespread and refined.

**3. Intelligent development of vehicles**

When driving in noisy environments, BCI technology can enhance key sound signals and filter out background noise through active noise control. For example, when the driver needs to concentrate, the system can reduce the volume of in-car music and amplify important sound signals such as sirens and horns, helping the driver better perceive the environment.

This active noise control technology not only enhances the driver's perception of important sound signals but also reduces the interference of environmental noise on the driver's attention, thereby improving the overall driving experience.

At the same time, BCI technology can dynamically adjust in-car audio settings based on the driver's personal preferences and attention state. When the system detects that the driver needs to concentrate, it can automatically adjust the volume of the sound system and play suitable background music to help the driver maintain a good attention state.

Through personalized audio settings, BCI technology can provide drivers with a more comfortable and safer driving environment, enhancing the overall driving experience. This kind of personalized adjustment can effectively improve the driver's auditory attention and reduce the risk of accidents caused by auditory distraction.

In summary, the importance of auditory perception in driving cannot be ignored. Through BCI technology, we can effectively enhance the driver's auditory attention and provide a safer and more comfortable driving experience. Current research applications also demonstrate the tremendous potential of this technology. With continuous progress and deepening application of BCI technology in improving driver attention, it will become an important technological force in improving the driving environment and enhancing driving safety.

## 10.3 The Impact of Brain–Computer Interface on Driver Spatial Attention

Spatial attention is a type of driving perceptual attention that combines visual attention and auditory attention. From a theoretical perspective, it can be regarded as the driver's ability to perceive, locate, and track objects and the environment within the driving environment. In daily driving, spatial attention is mainly used to judge the relative position and distance between the vehicle and surrounding objects (such as other vehicles, pedestrians, and obstacles) in order to make correct driving decisions.

Spatial attention not only involves the perception of static objects but also includes tracking and predicting dynamic targets, which is especially important in high-speed driving conditions such as highways and in three-dimensional navigation such as aviation. From the driver's perspective, the importance of spatial attention in driving is mainly reflected in three aspects.

### 10.3.1 *Deficits in Drivers' Spatial Attention and their Impact on Traffic Safety*

**1. Three-dimensional environmental perception:** The driver needs to accurately perceive the position of the vehicle or other traffic tools in three-dimensional space, as well as the relative distance to surrounding objects. This is crucial for avoiding collisions and ensuring safe driving.

**2. Decision-making ability:** Spatial attention directly affects the driver's decision-making ability. Operations such as lane changing, turning, parking, and decelerating all rely on accurate judgment of spatial information.

**3. Dynamic tracking:** When driving at high speeds or in complex road conditions, drivers need to continuously track the dynamic changes of surrounding vehicles to respond promptly.

In the current traffic driving process, many drivers lack sufficient spatial attention. This can lead to the following problems during driving.

### 10.3.2 *Enhancing Driver Spatial Attention via BCI-Integrated Spatial Mapping: From Technical Architecture to Multi-sensory Feedback*

**1. Driving Blind Spots:** If drivers do not maintain focused spatial attention during driving, some objects moving within the blind spot areas may be overlooked, which greatly increases driving risks and can lead to traffic accidents.

**2. Distance Judgment Errors:** Lack of spatial attention easily causes distance judgment errors, which is very dangerous when driving at high speeds or navigating in three-dimensional spaces. For example, during maritime navigation, inattentiveness can lead to distance misjudgments and consequently traffic accidents.

In July 2023, the Japan Coast Guard announced that a ship near the Miura Peninsula in Kanagawa Prefecture, Japan, collided with a whale, injuring multiple people on board. In July 2024, a similar incident occurred in New Hampshire, USA, involving a fishing boat and a whale. Although whale movements are indeed unpredictable, if drivers can promptly judge a safe distance and correct their course, such accidents can largely be avoided.

To enhance drivers' spatial attention, traffic technology researchers worldwide have actively conducted studies integrating BCI technology. This technology effectively helps drivers maintain high alertness regarding spatial orientation and improves driving safety. Currently, the primary method used globally to enhance spatial attention through BCI technology is spatial mapping technology.

Spatial mapping technology is a multidisciplinary field involving mathematics, computer science, and mixed reality technologies. Its main function is to capture scenes or objects in the physical world and present them as three-dimensional digital models in the virtual world, thereby enhancing perception of the scenes or objects. For example, the well-known digital twin technology is a spatial mapping technology that perfectly reproduces physical objects in digital space.

The application combined with BCI technology in the transportation field is a technology that integrates the driver's EEG signals with the environmental sensor data of the vehicle through a BCI system, generating a real-time three-dimensional spatial model through complex algorithms.

Currently, experts in related technical fields worldwide are deepening their research on this technology. In 2023, China's Yellow River News published an article titled "What Does the Yellow River Look Like in Digital Space Mapping? Let's Explore Together," which included a detailed analysis of spatial mapping technology.

Through analysis and review of this technical field, it can be concluded that currently, the spatial mapping technology integrating BCI and transportation technology mainly involves the following technical modules:

(1) *Data Acquisition*: The BCI system collects the driver's EEG signals in real time through EEG sensors, while the vehicle's sensors (such as LiDAR, ultrasonic sensors, and cameras) collect spatial data of the surrounding environment.
(2) *Signal Processing*: The collected EEG signals are preprocessed to filter out noise and interference, then feature signals related to spatial attention are extracted.
(3) *Feature Extraction*: Key features such as attention concentration and spatial perception ability are extracted from EEG signals using techniques such as time–frequency analysis and wavelet transform.
(4) *Denoising and Filtering*: Adaptive filters, principal component analysis (PCA), and other methods are used to remove noise and artifacts from the signals, improving signal quality.

(5) *Data Fusion*: The processed EEG signals are fused with environmental sensor data, and a real-time 3D spatial model is generated through multi-sensor data fusion algorithms.
(6) *Kalman Filtering*[2]: Using Kalman filters to fuse multi-sensor data, generating continuous and stable spatial models.
(7) *Bayesian Network*: Comprehensively analyzing multi-source data through Bayesian network models to provide more accurate environmental perception results.
(8) *Model Generation*: Based on fused data, generating 3D environmental models using computer vision and deep learning technologies to help drivers build spatial awareness of the surrounding environment.
(9) *Convolutional Neural Network (CNN)*[3]: Processing images and point cloud data with CNN to generate high-precision 3D models.
(10) *Recurrent Neural Network (RNN)*[4]: Combining time-series data to predict the motion trajectories of dynamic targets, improving the real-time performance and accuracy of the models.

Through the comprehensive application of the above technologies, BCI technology can integrate cutting-edge technologies such as big data, cloud computing, and AI to construct precise and real-time models of traffic movement in the digital space, thereby assisting drivers in enhancing spatial attention during the operation of vehicles.

For example, when vehicles are parking or driving on narrow roads, spatial mapping technology can provide real-time three-dimensional environmental information, helping drivers accurately judge the position of

---

[2]Kalman filtering is an algorithm that employs the state-space equations of a linear system to perform optimal estimation of the system's state based on observed input–output data. Since the observational data incorporates noise and disturbances from the system, this optimal estimation process can also be interpreted as a filtering operation.

[3]CNNs are a class of deep, feedforward neural networks that incorporate convolutional operations. As a representative algorithm of deep learning, CNNs possess the capability of representation learning. Their hierarchical architecture enables shift-invariant classification of input information, which is why they are also referred to as Shift-Invariant Artificial Neural Networks (SIANNs).

[4]Recursive Neural Network (RNN) is a type of Artificial Neural Network (ANN) with a tree-like hierarchical structure, in which network nodes recursively process input information according to their connectivity sequence. It is one of the algorithms under the umbrella of deep learning.

the vehicle body and the distance to obstacles through visual and tactile feedback, ensuring safe parking. When driving on complex urban roads, the system can provide dynamic information about surrounding vehicles and pedestrians through spatial mapping technology, helping drivers maintain a high level of spatial awareness.

Another example is during the process of ships docking and departing, spatial mapping technology can provide real-time three-dimensional environmental information, assisting ship pilots in accurately judging the position of the hull and the distance to docks, other ships, and obstacles, ensuring safe operations. When navigating narrow channels, the system can update channel information in real time, alerting pilots to avoid collisions and groundings.

Additionally, during takeoff and landing in air traffic, spatial mapping technology can provide real-time three-dimensional environmental information, helping pilots accurately judge the distance between the aircraft and the runway, other aircraft, and obstacles, ensuring safe operations. Moreover, when navigating complex airspace, the system can provide dynamic information about surrounding aircraft and obstacles through spatial mapping technology, helping pilots maintain a high level of spatial awareness.

Currently, the integration of BCI technology with other transportation technologies to develop spatial mapping technology has not yet achieved comprehensive practical application. However, in this direction, the fusion of BCI technology with transportation technology has already achieved various effects in enhancing drivers' spatial attention. These effects are mainly reflected in providing drivers with multi-sensory feedback based on BCI technology, thereby enhancing drivers' spatial perception. These multi-sensory feedbacks mainly include the following types.

### 10.3.3 *BCI-Enhanced Spatial Attention Systems: Multi-sensory Feedback Mechanisms and Technical Implementation*

**1. Tactile Feedback**

The tactile feedback system provides drivers with spatial orientation perception through vibrations on the steering wheel or seat. For example, when the driver approaches an obstacle or enters a blind spot, the system can alert them through vibrations, helping improve spatial attention.

The advantage of tactile feedback is that it does not occupy the driver's visual or auditory resources and can provide effective alerts without interfering with driving operations. When the driver approaches an obstacle while parking, the system can remind the driver through tactile stimulation, helping them accurately judge potential risks during traffic maneuvers.

The implementation of the tactile feedback system usually involves the following steps:

(1) *Vibration Signal Generation*: Based on environmental sensor data and the driver's EEG signals, the system generates vibration signals.
(2) *Pulse Code Modulation (PCM)*[5]: Encode vibration signals using pulse coding to improve signal accuracy.
(3) *Haptic Feedback Device Control*: Transmit vibration signals to haptic feedback devices (such as steering wheels or seats) to control their vibration intensity and frequency.
(4) *PWM Control*[6]: Use Pulse Width Modulation (PWM) technology to precisely control the operating state of vibration devices.
(5) *Vibration Feedback Transmission*: Haptic feedback devices transmit vibration signals to the driver, providing spatial orientation perception.
(6) *Multi-point Haptic Feedback*: Set multiple vibration points on control devices such as seats, steering wheels, and joysticks to provide more delicate haptic feedback.

Through this process, the haptic feedback system can provide effective spatial attention cues without interfering with driving operations, helping drivers enhance their perception of the surrounding environment.

**2. Auditory Feedback**

The auditory feedback system utilizes 3D sound technology to provide a sense of spatial orientation. For example, when there is an emergency behind the vehicle, the system can emit warning sounds through rear speakers to help the driver perceive and respond promptly.

---

[5]PCM is a method for converting continuously varying analog signals into digital signals through sampling, quantization, and encoding. The primary advantage of PCM is its high audio fidelity, while its main drawback is the large data size. PCM supports digital leased line services with data rates ranging from 2 Mbps to 155 Mbps, as well as other applications such as voice and image transmission, distance learning, and similar services.

[6]PWM is a control technique that involves modulating the width of pulses. By adjusting the duration of a series of pulses, it equivalently generates the desired waveform, including its shape and amplitude.

The application of auditory feedback is not limited to emergencies; it can also provide important spatial information during daily driving. Especially in high-speed and spatial driving environments, the system can use 3D sound effects to indicate changes in the traffic environment, helping the driver prepare in advance.

The implementation of the auditory feedback system generally involves the following steps:

(1) *Sound Signal Generation*: Generate 3D sound signals based on environmental sensor data and the driver's EEG signals.
(2) *Surround Sound Technology*: Use surround sound technology to generate spatial audio effects, providing a three-dimensional auditory experience.
(3) *Sound Processing*: Process the generated audio signals to ensure clarity and accurate localization of sounds.
(4) *Digital Signal Processing (DSP)*[7]: Optimize audio signals through DSP technology to enhance sound quality.
(5) *Sound Playback*: Play processed audio signals through the in-vehicle sound system to provide spatial orientation perception.
(6) *Multi-channel Audio*: Use multi-channel audio systems to achieve spatial localization of sound, improving the accuracy of auditory feedback.

Overall, the application of BCI technology in enhancing drivers' spatial attention demonstrates the great value of this technology, as well as various methods to improve spatial attention. It is believed that this technology will play an increasingly important role in highway, waterway, and aviation transportation, comprehensively improving the effectiveness of driver attention enhancement.

[7]DSP is a discipline that employs digital computation methods to perform operations such as signal transformation, filtering, detection, estimation, modulation, demodulation, and fast algorithms. DSP offers advantages including high precision, high reliability, programmable control, time-division multiplexing capability, and ease of integration. Its application areas are extensive.

# Chapter 11

# Application of Brain–Computer Interface in Intelligent Transportation Systems

## 11.1 Brain–Computer Interface and Traffic Information Sharing

In intelligent transportation systems, intelligent traffic management systems across the country face the daily tasks of receiving, sorting, and analyzing massive amounts of traffic information. The real-time nature and accuracy of information have become key factors for traffic safety and healthy development. With the acceleration of urbanization and the increase in the number of vehicles, enhancing the information processing capabilities of intelligent traffic management systems using modern technology has become an important task for building a strong transportation nation. As an innovative technology of the new era, brain–computer interface (BCI) technology is changing the way and effectiveness of traffic information sharing.

BCI technology itself is a highly intelligent information interaction technology that can not only reflect the driver's decisions and reactions in real time but also dynamically adjust according to real-time traffic data. Its application value in traffic information sharing is particularly prominent.

First, BCI technology makes the collection and transmission of traffic information more intelligent and efficient. This comprehensive information gathering and sharing enhances the perception capabilities of intelligent traffic management systems and provides a solid foundation for subsequent data processing and analysis.

Second, the application of BCI technology greatly improves the ability to process and analyze traffic information. Traditional traffic information processing often relies on preset rules and models, making it difficult to cope with complex and ever-changing traffic conditions. The BCI system, by introducing advanced algorithms such as deep learning and machine learning, can perform real-time analysis and processing of massive traffic data, thereby providing more accurate predictions and decision support.

Moreover, the application of BCI technology in traffic information sharing also holds significant social importance. It can provide timely warnings of potential dangers, reduce traffic accidents, and promote the development of vehicle-to-everything (V2X) and intelligent traffic systems, driving traffic management toward a more intelligent, collaborative, and sustainable direction.

Therefore, the application of BCI technology in traffic information sharing is not only an inevitable trend in technological development but also an effective means to address the challenges of intelligent transportation. We have every reason to believe that future traffic systems will develop to be more intelligent, efficient, and safe under the promotion of BCI technology.

### 11.1.1 *Brain–Computer Interface Technical Framework in Traffic Information Sharing*

Traffic information sharing is an important component of modern intelligent transportation systems, and BCI, as an emerging technology, is gradually changing the methods and efficiency of traffic information sharing. Currently, countries worldwide are striving to integrate BCI technology into intelligent transportation technology research and development, aiming to enrich the technical support for intelligent transportation development. Several countries have achieved remarkable results in traffic information sharing.

The Traffic Research Center at the Technical University of Munich (TUM), Germany, is currently conducting a project on the development and construction of intelligent transportation systems. This project integrates BCI technology and uses it to improve the efficiency and accuracy of traffic information sharing.

In recent years, the Traffic Research Center at the TUM has been dedicated to exploring the application of advanced technologies in the transportation field to address increasingly complex traffic problems. The intelligent transportation system project is an important research direction. In this project, the Traffic Research Center at TUM aims to leverage the characteristics of BCI technology to enhance driver state monitoring and traffic information sharing, optimizing traffic management and control.

According to information released in related academic reports, the technical system architecture design of this project includes four technical layers, which are as follows:

(1) *Perception Layer*: A high-precision dry-electrode EEG helmet is worn on the driver's head to collect brain electrical signals in real time. Meanwhile, various sensors such as high-definition cameras, LiDAR, and GPS are installed inside and outside the vehicle to obtain vehicle status and road environment information.
(2) *Transmission Layer*: Utilizes 5G networks to achieve high-speed and low-latency data transmission, ensuring that EEG signals and sensor data can be uploaded to cloud servers in real time for processing and analysis.
(3) *Platform Layer*: A powerful data management and analysis platform is established in the cloud, using distributed databases such as Apache Cassandra and big data processing frameworks like Apache Spark to achieve efficient data storage and processing. Through deep learning algorithms such as convolutional neural networks and long short-term memory networks, EEG signals are analyzed to extract key features such as driver attention and fatigue levels.
(4) *Application Layer*: Based on real-time traffic information and driver status, the intelligent transportation system can dynamically adjust traffic signals and traffic management strategies, and share information with other vehicles and traffic management centers through V2X technology.

Currently, the system is capable of real-time analysis of the driver's EEG signals to accurately determine their attention level and fatigue state. When the driver is distracted or fatigued, the system issues warnings and suggests resting or switching driving modes. By integrating EEG signals with vehicle status data, the system can predict traffic flow in real time and provide recommendations for dynamically adjusting the duration and phases of traffic signals.

The project underwent multiple field tests in Munich, and the results showed that the intelligent transportation system using BCI technology can significantly reduce traffic accidents and improve traffic flow efficiency. During the tests, the system successfully detected and handled multiple instances of driver distraction and fatigue, preventing potential traffic accidents.

The intelligent transportation system project at the Traffic Research Center of the TUM, Germany, realized driver status monitoring and real-time traffic information sharing through BCI technology, demonstrating the great potential of BCI technology in traffic information sharing. The success of this project provides valuable experience and reference for the future development of intelligent transportation systems.

From the research projects at the TUM, it can be seen that the technical framework of BCI technology in traffic information sharing is similar to its application logic in other traffic fields. The basic technical framework consists of five parts: sensing, signal processing, feature extraction, classification, and output control. These modules work together to realize the entire process from EEG signal acquisition to information transmission and sharing.

Among them, sensors are the front-end devices of the BCI system, responsible for collecting the driver's EEG signals. In practical applications, selecting high-sensitivity and low-noise sensors ensures the accuracy of signal acquisition. However, the collected EEG signals often contain considerable noise and interference, such as electromyographic signals and power frequency interference. The main task of the signal processing module is to filter and denoise the raw signals to extract effective EEG information. After signal processing, feature extraction is required for subsequent classification and analysis. There are various methods for feature extraction, mainly including time-domain features, frequency-domain features, and time-frequency domain features. Common time-domain features include mean, variance, and peak values; frequency-domain features use methods such as Fourier transform to extract energy information in different frequency bands; time–frequency domain features combine time and frequency information, with common methods including short-time Fourier transform and wavelet transform.

The extracted features also need to be classified by classifiers to identify the driver's intentions and states. After the classifier recognizes the driver's intentions, the output control module transmits the corresponding information to the traffic management system or vehicle control system. The design of the output control module needs to consider the real-time nature and reliability of the information to ensure that the driver's intentions are timely and accurately reflected on the traffic information sharing platform.

It can be seen that the technical framework of BCI technology applied in traffic information sharing is not complicated. Through the acquisition, processing, analysis, and transmission of EEG signals, information sharing between drivers and traffic management departments is realized. This mode of information sharing differs from traditional sharing, which is purely based on the analysis of vehicle and traffic status data. Instead, it integrates the driver's subjective state and experiential information, greatly enhancing the real-time nature, accuracy, and reliability of the information, thereby optimizing the real-time effects of traffic operation and management.

### 11.1.2 *Signal Processing Technology of Brain–Computer Interface in Traffic Information Collection*

The first step in traffic information sharing is the real-time collection of information, and the core application of BCI technology in this field lies in extracting useful information from EEG signals. This process involves complex signal processing techniques. The quality of signal processing technology directly affects the performance of the BCI system, thereby impacting the effectiveness of traffic information collection.

Signal processing in BCI technology has been a key research focus since the technology's inception, and the birth and development of this technology determine whether BCI can be practically implemented.

BCI Successfully Changes the Lives of People with Disabilities.

In 2014, at the World Cup opening ceremony in São Paulo, Brazil, 29-year-old paraplegic patient J. Pinto kicked off the first ball using an exoskeleton controlled by brain signals. This event attracted worldwide attention and brought extensive focus to BCI technology. BCI technology establishes a direct connection between the brain and external devices, enabling real-time transmission of brain intentions and providing new rehabilitation hope for paraplegic patients. Since then, how to convert brain intentions into commands for controlling external devices in real time and effectively has become a key issue in the development of BCI technology.

Soon after, the research team led by Fengzhen Tang at the National Key Laboratory of Robotics in China conducted a series of studies on decoding scalp EEG signals, aiming to improve the efficiency and accuracy of EEG signal decoding. They proposed a brain signal decoding method based on Log-Euclidean Metric Riemannian Geometry, which not only enhanced decoding efficiency but also ensured the precision of signal decoding.

The researchers represented EEG signals as covariance matrices, transforming them from flat Euclidean space to curved symmetric positive-definite Riemannian space, and applied the Log-Euclidean Metric (LEM) for signal processing, as shown in Figure 11.1.

This method extends the Generalized Learning Vector Quantization (GLVQ) to Riemannian space, forming the GLVQ method based on Log-Euclidean distance (GLVQ-LEM). The researchers also introduced a Log-Euclidean Metric learning approach to learn a function that maps the original manifold to a more separable Riemannian submanifold, thereby improving the accuracy and efficiency of signal processing, as shown in Figure 11.2.

The study validated the proposed method on multiple datasets, achieving excellent results in both recognition speed and accuracy. Compared to EEG signal decoding methods based on affine-invariant

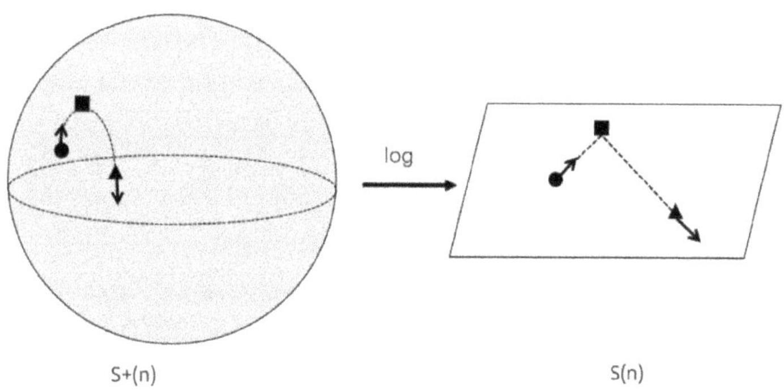


**Figure 11.1.** Schematic diagram of the generalized learning vector quantization method based on log-Euclidean metric.

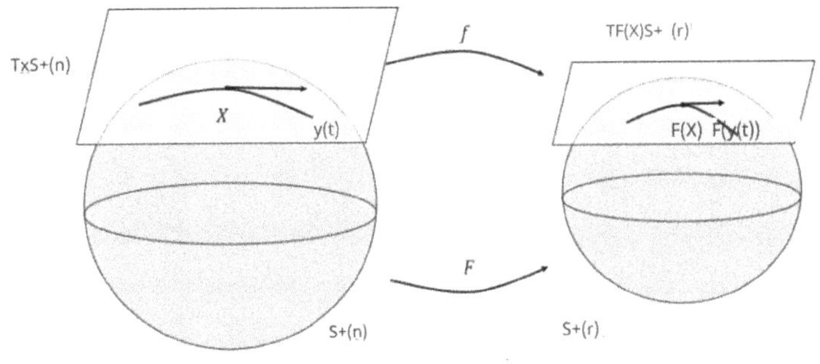


**Figure 11.2.** Log-Euclidean metric learning framework.

Riemannian metrics, the new method improved recognition speed by approximately four times. In tests on this dataset, the proposed method's recognition accuracy surpassed that of the competition's first-place result. The new method significantly accelerated brain signal decoding and effectively reduced the latency in executing commands within the BCI system.

This study is of great significance for advancing the practical application of brain–computer interfaces (BCIs) in motor rehabilitation for paralyzed patients. Through efficient EEG signal decoding technology, paralyzed patients can control external devices more quickly for motor rehabilitation training.

Nowadays, the information collection and processing technologies of BCIs have been comprehensively developed and deeply applied in various fields. However, from the perspective of traffic information collection, several important technical challenges still need to be addressed.

1. **High real-time requirements:** Effective collection and transmission of traffic information require a high degree of real-time performance to promptly adjust traffic signals and management strategies. BCI technology needs to utilize efficient signal processing algorithms to decode drivers' intentions and emotions in real time, thereby achieving faster traffic information responses.

2. **Multi-source information fusion:** Traffic information includes not only vehicle position and speed but also the driver's status information. BCI technology needs to integrate traditional traffic data to provide a more comprehensive traffic information collection solution.

**3. Effective integration with big data technology:** With the development of intelligent transportation systems, the volume of traffic information data has exploded. Although BCI technology can effectively process traffic data through advanced signal processing and machine learning algorithms, it still needs to be effectively integrated with big data, cloud computing, and other technologies to fully realize its potential when facing massive amounts of information.

By overcoming the above three challenges, the application effectiveness of BCI technology can be significantly enhanced, thereby playing a prominent role in signal processing for traffic information collection and providing greater impetus for traffic development.

### 11.1.3 *Advanced Algorithms of Brain–Computer Interfaces in Traffic Information Processing and Analysis*

BCI system provides a rich data source for intelligent transportation systems by collecting and transmitting electroencephalogram signals and traffic information. Effective processing and analysis of these data rely on advanced algorithms and technologies. The application of advanced algorithms in BCI systems can not only improve the accuracy and efficiency of traffic information processing but also enable the prediction and optimization of traffic conditions.

BCI Research Project at Daegu Gyeongbuk Institute of Science and Technology (DGIST), South Korea.

In recent years, the research team at DGIST, South Korea, has been dedicated to addressing issues related to neural recording data processing through machine learning (ML) techniques. They have achieved a research breakthrough in reconstructing high-frequency neuronal spikes from low-frequency signals obtained by secondary sampling based on ML. This research result significantly enhances the algorithmic capabilities of BCI technology in information processing and analysis, providing new possibilities for the application of BCI technology in traffic information processing.

Inspired by the equivalence between high-frequency recovery and super-resolution in image processing, the researchers applied the Transformer ML model to neuronal data processing. Specifically, they proposed a spike reconstruction model named Spk-Recon, which can reconstruct high-frequency spike information from significantly downsampled low-frequency neuronal signals, as shown in Figure 11.3.

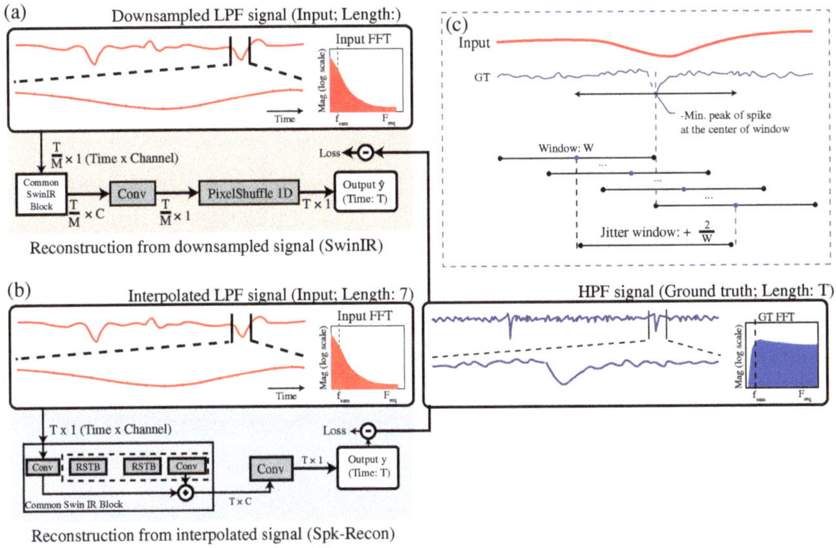


**Figure 11.3.** Spk-Recon model architecture and training diagram.

The core of this model lies in utilizing low-pass filtering and signal interpolation techniques to input low-sampling-rate data into the ML model and recover high-time-resolution high-frequency spike signals through ML algorithms. This technology can realize various applications in traffic information processing and analysis.

Considering the demands of modern intelligent transportation development, it is clear that this technology can be fully applied in traffic information analysis and processing. Through BCI technology and the Spk-Recon model, traffic management systems can monitor drivers' EEG signals in real time, detect their fatigue states and emotional changes, and accordingly adjust traffic signals and management strategies in a timely manner to improve traffic flow and safety. By combining traditional traffic data with EEG signals, BCI technology can provide a more comprehensive traffic information collection and analysis solution, enhancing the intelligence level of traffic management. The Spk-Recon model can also leverage its efficient signal processing and machine learning algorithms to handle massive traffic data and extract useful information, improving the efficiency of traffic information processing and analysis.

Additionally, researchers at DGIST applied the Spk-Recon model to multi-channel neural recording datasets from *in vitro* hippocampal neurons and *in vivo* mouse brains. Through traditional timing and waveform quantitative spike analysis, it was demonstrated that the model can accurately reconstruct spike signals from significantly downsampled low-frequency neuronal signals, with a spike occurrence hit rate close to 0.8–0.9 and spike sorting clustering accuracy exceeding 96%.

The model records low-frequency band signals at a uniformly lower sampling rate and recovers high-sampling-rate spike signals through ML software. This means that in various commercial or custom multi-channel neural recording systems, signal acquisition with reduced data volume can be achieved without additional hardware modifications.

It can be seen that machine learning-based spike reconstruction technology is not only significant in the field of neuroscience but also provides new solutions for traffic information processing and analysis. Through this technology, traffic management systems can achieve more efficient real-time monitoring and intelligent management, improve traffic flow and safety, and promote the development of intelligent transportation systems.

### 11.1.4 *System Integration of Brain–Computer Interface in Traffic Information Sharing Platforms*

The application of BCI technology enables traffic information sharing platforms to collect, process, and transmit traffic data more efficiently, achieving the optimization and management of intelligent transportation systems. System integration is the key to ensuring the efficient operation of these functions. Through reasonable system architecture design and advanced technical means, organic integration of various modules can be realized, enhancing the overall system performance. Next, let us understand how BCI technology is integrated into intelligent traffic information sharing platforms.

The application logic of BCI technology has been analyzed many times before, so it will not be repeated here. We directly analyze from the design perspective of BCI technology in traffic information sharing platforms, focusing on how this technology can be applied to traffic information systems within the platform and what role it can play.

1. **Intelligent Traffic Management System:** Through BCI technology, the intelligent traffic management system can integrate multi-source data collection, real-time data processing, and deep analysis functions, providing comprehensive traffic management and decision support. By monitoring

and analyzing traffic flow data in real time, the intelligent traffic management system can dynamically adjust traffic signal durations, optimize traffic flow, and improve road throughput efficiency. For example, during peak traffic periods, the system can adjust signal durations and phases based on real-time traffic flow data to alleviate congestion.

**2. Intelligent Driving System:** The intelligent driving system, through BCI technology, monitors the driver's status and intentions in real time, combined with vehicle sensor data, to achieve intelligent driving assistance and autonomous driving functions. The system can perceive the driver's attention and fatigue levels in real time, providing corresponding assistance in decision-making and control.

**3. Traffic Information Service Platform:** Through BCI technology, the traffic information service platform can integrate real-time collection and processing of multi-source data, providing comprehensive traffic information services. The platform can publish real-time traffic flow, road conditions, and accident information, helping drivers choose the best routes and avoid congestion.

In summary, system integration of BCIs in traffic information sharing platforms, through reasonable architecture design and advanced technical means, can achieve efficient collection, processing, and transmission of multi-source data, enhancing the intelligence level of traffic management and decision-making. This allows traffic information to be collected and analyzed more effectively, while enabling each piece of information to play its role, providing better decision-making effects for traffic operation and optimization.

## 11.2 Brain–Computer Interface and Traffic Congestion Management

Transportation development has promoted the popularization of various types of vehicles, while the rapid increase in vehicles has triggered traffic congestion problems, which in turn restrict transportation development. Nowadays, this development issue has become a severe challenge faced by major cities worldwide, and solving these problems is the primary task for the healthy development of transportation.

Traditional traffic management methods, including traffic signal control, road expansion, and public transportation optimization, although alleviating congestion to some extent, have increasingly revealed their limitations. Existing methods often rely on static and delayed data, making it difficult to respond to complex and dynamic traffic conditions in real time.

Against this background, BCI technology, with its unique advantages, has become a new hope for solving traffic congestion problems. From academic and technical perspectives, the application prospects of BCI technology are broad. Currently, multiple studies have explored the potential applications of BCI in traffic signal control, autonomous driving, and traffic flow prediction. BCI-based traffic signal control systems can dynamically adjust signal timing according to real-time collected traffic data to optimize traffic flow. Moreover, BCI technology can be combined with big data analytics and artificial intelligence algorithms to improve the accuracy and reliability of traffic flow forecasting.

In summary, BCI technology shows great potential in the field of traffic congestion management. Through continuous technological innovation and optimization, BCI technology is expected to bring new breakthroughs to the development of intelligent transportation.

### 11.2.1 *The Current Situation and Challenges of Traffic Congestion in China*

Traffic congestion not only affects the travel efficiency of the public but also has profound negative impacts on urban economic development, environmental protection, and social harmony. However, as of 2024, traffic congestion remains a pressing issue that needs to be addressed urgently in China's transportation development.

According to the "2023 Annual China Urban Traffic Report" released by Baidu Maps, as shown in Figure 11.4, the top five cities with the longest average commuting times in 2023 were Beijing, Shanghai, Nanjing, Tianjin, and Dalian, with Beijing averaging 44.47 minutes per commute. The top five cities with the longest average commuting distances in 2023 were Beijing, Chongqing, Shanghai, Chengdu, and Changchun, with Beijing averaging 12.53 km per commute. Beijing remains the most congested city, ranking first, while Shanghai's congestion coefficient and commuting time have increased significantly. The top 10 cities with peak-hour traffic congestion in 2023 were Beijing, Chongqing, Guangzhou, Shanghai, Wuhan, Changchun, Nanjing, Xi'an, Shenyang, and Lanzhou. Among them, Beijing's congestion index rose by 20.13% year-on-year, with an actual peak commuting speed of 24.26 km/h and a congestion index of 2.125, far exceeding other cities. Lanzhou was the city with the

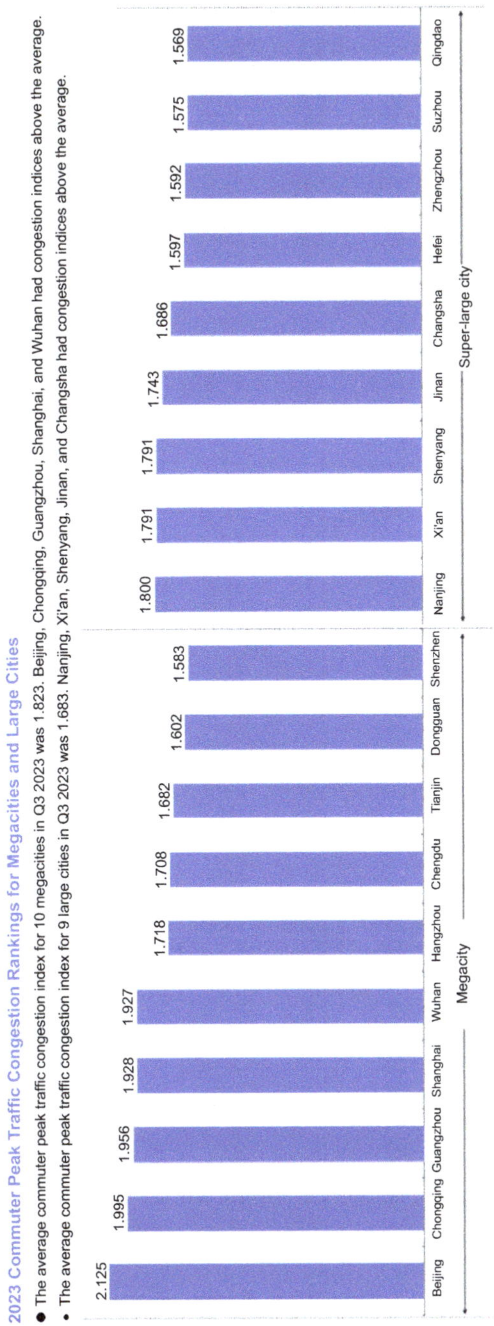


**Figure 11.4.** 2023 Annual China Urban Traffic Report.

fastest rise in congestion ranking, climbing 36 places. These data indicate that urban congestion not only hinders healthy development but also significantly reduces the quality of life for the public.

In fact, China's transportation development speed is among the fastest globally, especially with the advancement of transportation technology greatly improving the quality of transportation development. However, the problem of traffic congestion has not been fundamentally resolved. Based on the current status of China's transportation development, four main causes of traffic congestion can be identified.

#### 11.2.1.1 *Structural drivers of urban traffic congestion: vehicle proliferation, resource scarcity, and systemic limitations*

**1. Rapid Increase in Vehicle Numbers:** The rapid increase in the number of vehicles is one of the important factors contributing to traffic congestion. With economic development and the improvement of residents' living standards, the ownership of private cars has grown rapidly. In many cities, private cars have become the main means of daily travel for citizens. However, urban road resources are limited, and the rapid growth in the number of vehicles has exceeded the road capacity, leading to increasingly severe traffic congestion. According to statistics, China's vehicle ownership has been rising year by year in recent years. By the end of 2023, the total number of motor vehicles nationwide exceeded 400 million, with car ownership reaching 270 million, placing enormous pressure on urban transportation.

**2. Limited Road Resources:** Limited urban road resources are also an important cause of traffic congestion. Although cities continuously expand and build new roads, the supply of road resources appears insufficient compared to the rapidly growing traffic demand. Especially in the urban core areas, road resources are even more strained. Road construction requires a large amount of land resources, while urban land resources are inherently limited, leading to increased difficulty and cost in road expansion. In addition, road construction must also consider factors such as environmental protection and urban planning, all of which restrict the capacity for road resource expansion.

**3. Imperfect Public Transportation System:** The imperfection of the public transportation system is another important cause of traffic congestion. In some cities, the public transportation system has not yet reached the ideal state of wide coverage, convenient transfers, and efficient operation, causing citizens to prefer private car travel. Moreover, aging public transportation facilities and insufficient maintenance also affect service quality and attractiveness. For example, limited coverage of bus and subway lines, low frequency of service, and poor punctuality are all issues that the public transportation system urgently needs to address.

**4. Unscientific Traffic Management:** Unscientific traffic management is another cause of traffic congestion. Existing traffic management methods often lack scientific and systematic approaches and fail to fully utilize modern technological means for intelligent management. For example, the setting and timing of traffic signals often lack real-time data support and cannot dynamically adjust according to actual traffic flow, resulting in traffic signals not achieving their maximum effectiveness. In addition, traffic management departments often fail to fully consider traffic flow trends and future development needs when formulating traffic policies and plans, causing traffic management measures to lag behind actual demands.

In recent years, although China's traffic departments have achieved certain results in solving traffic congestion problems, the limitations of existing traffic management methods have become increasingly apparent. Traditional traffic management methods mainly include traffic signal control, road expansion, and public transportation optimization, but these methods have many shortcomings in practical application and have not fundamentally solved the problem of traffic congestion.

#### 11.2.1.2 *Limitations of conventional traffic management strategies: From signal control to policy planning*

**1. Limitations of Traffic Signal Control:** Traffic signal control is one of the most common traffic management methods, but it has obvious limitations in solving traffic congestion. Most existing traffic signal control systems are based on fixed timing or simple traffic flow sensing and cannot dynamically adjust signal timing according to real-time traffic conditions. This leads to difficulties in flexibly responding to traffic flow fluctuations, often resulting in unreasonable signal settings and uneven distribution of red and green light durations, further exacerbating traffic congestion. For example, during peak hours, the traffic signal settings at some intersections cannot meet actual demands, causing vehicles to queue for long periods and resulting in low traffic efficiency.

**2. Limitations of Road Expansion:** Road expansion is another common method to alleviate traffic congestion, but its effects are often temporary. With the continuous advancement of urbanization, the demand for road expansion is increasing day by day. However, urban land resources are limited, and the cost of road expansion is high with a long cycle, often making it difficult to meet the rapidly growing traffic demand. In addition, road expansion faces many challenges, such as environmental protection and relocation issues, which increase the difficulty and complexity of road expansion. More importantly, road expansion cannot fundamentally solve the traffic congestion problem, because the increase in the number of vehicles often quickly fills the added road capacity, causing congestion to reoccur.

**3. Limitations of Public Transportation Optimization:** Optimizing public transportation systems is one of the important means to alleviate traffic congestion, but there are many shortcomings in its practical application. First, the construction and maintenance of public transportation facilities require substantial financial investment, and in some cities, due to budget constraints, the construction and optimization of public transportation systems often fail to keep pace with the growing demand. Second, there are many problems in the operation and management of public transportation systems, such as low service frequency, poor punctuality, and inconvenient transfers, which reduce citizens' willingness to use public transportation. In addition, the coverage of public transportation systems is limited, and residents in some suburban and remote areas have difficulty accessing convenient public transportation services. This further increases the use of private cars, exacerbating traffic congestion problems.

**4. Limitations of Traffic Policies and Planning:** Traffic policies and planning are key components in solving traffic congestion problems, but existing traffic policies and planning often lack foresight and scientific rigor. On one hand, the formulation of traffic policies often lags behind actual needs and fails to respond promptly to changes in traffic flow trends. For example, some cities, when formulating traffic restriction policies, do not fully consider vehicle travel demand and road capacity, resulting in poor implementation effects of these policies. On the other hand, the scientific and systematic nature of traffic planning is insufficient, and modern technological means are not fully utilized for traffic flow prediction and management. For instance, traffic planning departments often lack comprehensive analysis of traffic flow data when designing and arranging roads, leading to unreasonable road layouts and poor traffic flow.

In summary, solving traffic congestion problems requires not only scientific and reasonable traffic management methods but also the comprehensive application of modern technological means to continuously optimize and innovate traffic management approaches. As a cutting-edge technology, BCI technology has begun to demonstrate great application potential in traffic congestion management.

### 11.2.2 *Application Strategies of Brain–Computer Interface Technology in Solving Traffic Congestion*

As a cutting-edge technology, BCI technology provides new ideas and methods for solving traffic congestion problems. In response to the current status and challenges of traffic congestion analyzed earlier, BCI technology can play an important role in traffic management, with applications reflected in big data processing, algorithm optimization, system integration, and compatibility.

**1. Real-time Traffic Data Collection and Processing**

BCI technology uses EEG signal acquisition devices, such as EEG helmets, to monitor drivers' brainwave activity in real time. These devices can be installed on the driver's seat or other locations inside the vehicle without affecting the driver's normal operation. Through wireless transmission technology, these brainwave signals can be transmitted in real time to the traffic management center for data analysis and processing.

At the traffic management center, the collected brainwave signals are integrated with traditional traffic data. Using big data analysis technology, traffic conditions can be monitored and analyzed in real time, including traffic flow, vehicle speed, and road congestion. At the same time, BCI technology can also monitor the driver's cognitive state in real time, determining whether their attention is focused or if there is fatigue driving. Through data fusion and analysis, the traffic management system can promptly detect abnormal traffic conditions and take corresponding measures for intervention.

Based on real-time collected and analyzed data, BCI technology can optimize traffic signal control systems. Traditional traffic signal control systems often use fixed timing or simple sensor-based control, which struggle to cope with complex and variable traffic conditions. BCI technology can dynamically adjust signal timing according to real-time data, improving the efficiency of traffic flow.

**2. Algorithm Optimization and Big Data Processing**

The processing and analysis of traffic data are key steps to achieving intelligent traffic management. BCI technology, combined with big data processing and artificial intelligence algorithms, can enhance the accuracy and reliability of traffic flow prediction, thereby more effectively addressing traffic congestion issues.

Through big data processing and artificial intelligence algorithms, BCI technology can accurately predict traffic flow. Traffic flow prediction refers to forecasting the trend of traffic volume changes over a future period based on historical and real-time data. By real-time monitoring of drivers' cognitive states, BCI technology can obtain more traffic behavior data, further improving the accuracy of traffic flow prediction.

Based on traffic flow prediction, BCI technology can further optimize traffic management algorithms. Traffic optimization algorithms include traffic signal timing optimization, traffic stream optimization, and vehicle dispatch optimization. With BCI technology, traffic management systems can obtain real-time data on drivers' cognitive states and traffic conditions, dynamically adjusting traffic optimization algorithms.

BCI technology combined with big data processing and artificial intelligence algorithms can also perform deep mining of traffic data to support traffic management decision-making. Data mining refers to analyzing large amounts of traffic data to discover hidden traffic patterns and problems and propose corresponding solutions. For example, BCI technology can analyze traffic accident data to identify high-incidence accident locations and causes, providing a basis for traffic management departments to formulate traffic safety measures.

**3. System Integration and Compatibility**

The application of BCI technology requires seamless integration with existing traffic management systems, which raises higher demands for system integration and compatibility. Through reasonable system integration schemes, the advantages of BCI technology can be fully utilized to improve the intelligence level of traffic management.

During system integration, existing traffic management systems need to be redesigned to ensure seamless interfacing with BCI technology. System architecture design includes both hardware and software architecture. Hardware architecture design mainly involves the installation and configuration of BCI devices to ensure stable and reliable operation. Software architecture design mainly includes the design of data collection, transmission, processing, and analysis modules to ensure smooth and efficient data flow.

To achieve seamless interfacing between BCI technology and existing traffic management systems, unified data interfaces and communication protocols need to be designed. Data interfaces refer to data exchange channels between different systems, and communication protocols refer to the rules and standards for data transmission. By designing unified data interfaces and communication protocols, it can be ensured that data collected by BCI devices can be smoothly transmitted to traffic management systems for processing and analysis.

After the system integration is completed, comprehensive system compatibility testing is required to ensure that the BCI technology can operate stably and reliably with the existing traffic management system. System compatibility testing includes two aspects: hardware compatibility testing and software compatibility testing. Hardware compatibility testing mainly involves testing the compatibility between the BCI devices and the existing hardware devices to ensure their normal operation. Software compatibility testing mainly includes testing the compatibility of data collection, transmission, processing, and analysis modules to ensure smooth and efficient data flow.

In summary, BCI technology demonstrates tremendous application potential in traffic congestion management. Through its innovation and optimization in the field of intelligent transportation, BCI technology is expected to bring new breakthroughs to the development of smart transportation, helping achieve a more efficient, safe, and intelligent traffic management system.

## 11.3 Brain–Computer Interface and Traffic Planning and Design

Besides the issue of traffic congestion, problems such as unreasonable road planning and imperfect public transportation system design have increasingly become prominent nowadays, which largely affect the operational efficiency of cities and the quality of life of residents. In this regard, the value of BCI technology should not be overlooked, as it can also play a significant role in improvement.

### 11.3.1 *Current Status and Challenges of Traffic Planning and Design*

Currently, China's transportation development has shown a trend toward intelligence, and the construction of a comprehensive three-dimensional transportation network nationwide is becoming more comprehensive. In this process, traffic planning and design play an important role. From the current situation of China's transportation development, traffic planning and design mainly include three aspects: urban traffic planning, road design, and public transportation system design.

### 11.3.1.1 *Shortcomings in current transportation planning and design: from forecasting to implementation*

**1. Urban Traffic Planning:** Urban traffic planning is an important part of the overall urban planning, aiming to optimize the traffic network structure through reasonable layout of transportation infrastructure, meet urban traffic demands, and improve the overall efficiency of the traffic system. Urban traffic planning includes road network planning, public transportation system planning, and non-motorized and pedestrian system planning. In modern cities, traffic planning must consider not only traffic flow and traffic efficiency but also environmental protection, energy conservation, and sustainable development.

**2. Road Design:** Road design is a crucial part of traffic planning, involving multiple aspects such as road layout, width, slope, and pavement materials. Scientific and reasonable road design can enhance the traffic flow capacity and reduce the occurrence of traffic accidents. Currently, road design is gradually moving toward intelligence and refinement by introducing advanced traffic monitoring and management systems to achieve real-time monitoring and dynamic management of road traffic.

**3. Public Transportation System Design:** The design of public transportation systems aims to build an efficient, convenient, and economical public transit network to meet the travel needs of urban residents. Public transportation system design includes bus route planning, subway line planning, and the construction of public transportation hubs. A well-developed public transportation system can not only alleviate urban road traffic pressure but also reduce energy consumption and environmental pollution, thereby improving the overall operational efficiency of the city.

Although current transportation planning and design have achieved certain results and have largely alleviated urban traffic problems, they still do not fully meet the goals of high-quality and high-speed transportation development. There remain some deficiencies and challenges that need to be addressed promptly. These challenges mainly manifest in inaccurate traffic flow forecasting, unreasonable road layouts, and imperfect public transportation system design.

#### 11.3.1.2 *Multi-dimensional challenges in traffic planning and design: forecasting, layout, systems, and implementation*

**1. Inaccurate Traffic Flow Forecasting:** Traffic flow forecasting is the foundation of transportation planning and design, but current traffic flow forecasting methods still have significant inaccuracies. Traditional traffic flow forecasting mainly relies on historical data and empirical judgment, making it difficult to fully consider the dynamic changes and complexity of traffic flow. Especially during peak traffic periods and major events, traffic flow often experiences large fluctuations, which traditional forecasting methods struggle to handle. This results in transportation planning and design being unable to accurately estimate future traffic demand, affecting the overall efficiency of the transportation system.

**2. Unreasonable Road Layout:** Unreasonable road layout is one of the main causes of traffic congestion. In some cities, transportation planning lacks scientific and forward-looking approaches, leading to unreasonable road layouts and poor traffic flow. For example, issues such as overly complex road network structures, poor connections between main roads and branch roads, and mismatches between road width and traffic volume all affect the capacity of traffic flow. In addition, road construction often fails to fully consider future traffic development needs, causing newly built roads to quickly become congested.

**3. Public Transportation System Design Needs Improvement:** Imperfect public transportation system design is also an important challenge faced by current transportation planning and design. At present, some cities in China have problems such as unreasonable route planning, inconvenient transfers, and low operational efficiency in their public transportation systems. Specifically, bus routes have limited coverage, with insufficient public transportation services in some residential and commercial areas; subway line layouts are unreasonable, causing inconvenient transfers and long travel times for passengers; public transportation hub designs are unscientific, with incomplete transfer facilities, negatively impacting passengers' travel experience. These issues not only reduce the attractiveness of public transportation but also increase traffic pressure on urban roads.

**4. Environmental and Social Impacts:** Transportation planning and design also need to consider environmental protection and social impacts. Road construction and expansion often require large amounts of land resources, which may lead to ecological damage and land resource waste. Furthermore, transportation planning and design must pay attention to social equity to ensure that all residents can enjoy convenient transportation services. Currently, some cities have not fully considered environmental protection and social impacts in their transportation planning and design, resulting in ecological degradation and intensified social conflicts.

**5. Technology and Management Need Improvement:** With the development of technology, traffic management technology is continuously advancing. However, in practical applications, traffic planning and design often lag behind technological progress. For example, the application of intelligent traffic management systems still faces many technical and managerial bottlenecks, making it difficult to fully realize their potential. In addition, traffic management departments often lack systematic and scientific approaches when formulating and implementing traffic plans, making it difficult to establish effective management mechanisms.

**6. Funding and Policy Constraints:** Traffic planning and design require substantial financial investment, and insufficient funding is one of the main factors restricting the construction and optimization of traffic infrastructure. Due to budgetary constraints, many cities find it difficult to smoothly implement traffic planning and design projects. Moreover, the lagging and imperfect traffic policies are also important factors affecting the effectiveness of traffic planning and design.

Therefore, there is still room for improvement in current traffic planning and design. Utilizing modern technologies such as big data, cloud computing, BCIs, and AI to comprehensively address the aforementioned issues and enhance the effectiveness of traffic planning and design is indeed an effective development approach.

### 11.3.2 *Application of Brain–Computer Interface Technology in Traffic Planning and Design*

Currently, China's traffic management departments are continuously improving traffic development outcomes nationwide and have provided extensive guidance for improvements in planning and design. Furthermore, the State Council issued "14th Five-Year Plan for the Development of a Modern Comprehensive Transportation System" explicitly emphasizes adhering to innovation as the core, enhancing development momentum, promoting the empowerment of new technologies to improve the quality and efficiency of transportation development, and strengthening the resilience of the comprehensive transportation system. Most of these goals are also reflected in traffic planning and design.

By combining the characteristics of BCI technology and its related applications in the field of intelligent transportation development, it can be seen that it can play a role in the following aspects to effectively address the existing problems and challenges in traffic planning and design mentioned above.

**1. Road Design and Layout Optimization**

Unreasonable road layout is one of the main causes of traffic congestion. BCI technology can provide more scientific data support for road design by real-time collection and analysis of the behavior data of traffic participants. For example, by analyzing the behavioral responses of drivers and pedestrians under different road conditions, BCI systems can optimize road width, curvature, and signage settings to improve road capacity and safety.

Intelligent road layout based on BCI technology can also effectively address the problem of unreasonable road layouts. BCI systems dynamically adjust road layouts and traffic flow lines by real-time monitoring of traffic volume and road usage.

**2. Public Transportation System Design Optimization**

Imperfect design of public transportation systems is also a major challenge in current transportation planning and design. BCI technology can provide scientific basis for public transportation route planning by real-time collection and analysis of passengers' EEG signals. For example, BCI technology can support the optimization of public transit transfers by monitoring and analyzing passengers' transfer behaviors in real time. By analyzing passengers' EEG signals during the transfer process, the BCI system can optimize the design of transfer stations and transfer routes, reduce passengers' transfer time, and improve the operational efficiency of the public transportation system.

**3. Comprehensive Consideration of Environmental and Social Impacts**

Transportation planning and design need to take environmental protection into account. BCI technology can analyze the EEG signals reflecting users' attitudes toward the environment of different road sections, thereby identifying public goals for road environment construction. Subsequently, it can optimize road planning and design strategies to enhance the construction and design effectiveness of the transportation environment.

Moreover, transportation planning and design must pay attention to social equity, ensuring that all residents can enjoy convenient transportation services. BCI technology can optimize transportation planning and design by real-time monitoring and analyzing the behaviors and needs of transportation participants, thereby enhancing the inclusiveness of the transportation system. For example, transportation authorities can use BCI technology to analyze the travel demands of different population groups, and then optimize the layout of public transit routes and transfer facilities to improve the fairness and accessibility of transportation services.

It can be seen that by integrating modern technological means to optimize transportation planning and design, the operational efficiency of urban transportation systems can be effectively improved, promoting sustainable urban development. The uniqueness of BCI technology can enhance the alignment between transportation planning and public needs, bringing new breakthroughs to intelligent transportation development and new possibilities for smart transportation advancement.

# Chapter 12

# Insights of Brain–Computer Interface Technology on Urban Transportation Planning

## 12.1 The Impact of Brain–Computer Interface on Traffic Demand Forecasting

The core of China's transportation development is not only to build world-class transportation but also to build transportation that satisfies the people. This is also the core concept of China's transportation construction, which centers on the people to coordinate the high-quality development of integrated transportation.

What is transportation that satisfies the people? It means building a high-quality transportation system that meets the diverse travel and life needs of the people. In this field, brain–computer interface (BCI) technology holds more influence compared to other transportation technologies.

In today's rapidly advancing wave of technology, BCI technology is gradually becoming an emerging tool for solving traffic problems with its groundbreaking application potential. By capturing and analyzing human brain activity signals, BCI technology enables direct communication between humans and computers. This not only opens a new window for studying human behavior but also injects unprecedented accuracy and dynamic responsiveness into traffic demand forecasting. Traditional traffic demand forecasting often relies on historical data and static models, making it difficult to cope with the ever-changing traffic conditions in modern cities. The introduction of BCI technology will provide traffic planners with real-time, precise behavioral data, making traffic demand forecasting more scientific and forward-looking. Next, we analyze in detail how BCI technology redefines traffic demand forecasting and provides strong support for the construction of intelligent transportation systems.

### 12.1.1 *Traffic Demand Forecasting is the Weather Vane of Traffic Development*

Traffic demand forecasting, as a core part of traffic planning and management, is not only the weather vane of traffic development but also a key tool guiding the formulation and decision-making of traffic policies. The construction of China's comprehensive three-dimensional transportation system is based on meeting demands derived from traffic demand forecasting. Accurate traffic demand forecasting can help cities plan traffic layouts more reasonably and improve the effectiveness of traffic development. In the following, we start from the role of traffic demand forecasting to analyze its importance and how BCI technology can play a prominent role in this field.

The efficient operation of the traffic system is directly related to the city's economic development and the convenience of residents' travel. Through accurate traffic demand forecasting, traffic can be reasonably planned, cities can develop healthily, and the quality of life for the public can be improved.

#### 1. Reasonable Traffic Planning

The primary role of traffic demand forecasting is to clarify the specific travel demands of urban residents and the needs of urban development, and based on this, to design and optimize traffic construction.

(1) *Road Network Planning*: Traffic demand forecasting can provide detailed traffic flow data to help planning departments reasonably arrange the road network. By forecasting future traffic demands, it is possible to foresee which areas will become traffic hotspots, thereby planning and constructing corresponding road infrastructure in advance to avoid future traffic bottlenecks. For example, in emerging development zones, accurate traffic demand forecasting can reasonably plan the layout of main roads and branch roads to ensure smooth traffic flow.

(2) *Traffic Signal Optimization*: Traffic signal timing is an important factor affecting road capacity. Through traffic demand forecasting, traffic signal timings can be dynamically adjusted to optimize traffic flow distribution and improve road traffic efficiency. During peak traffic periods, with real-time traffic demand forecasting, traffic management departments can promptly adjust signal timings to reduce vehicle waiting times and alleviate traffic congestion.

(3) *Public Transportation System Optimization*: Public transportation systems are an important means to alleviate urban traffic pressure. Through traffic demand forecasting, the layout of bus routes and subway lines can be optimized to improve the coverage and service quality of public transportation. For example, by analyzing traffic demand forecast data, bus schedules and routes can be adjusted to ensure that the public transportation system meets citizens' travel needs, reduces private car usage, and decreases traffic congestion.

**2. Supporting Urban Planning and Development**

Traffic demand forecasting not only is crucial for the operational efficiency of transportation systems but also has a significant impact on the overall urban planning and development. Accurate traffic demand forecasts can provide a scientific basis for urban planning and promote sustainable urban development:

(1) *Scientific Urban Planning*: Traffic demand forecasting can provide detailed data on traffic flow and travel patterns, assisting urban planning departments in formulating scientific urban development plans. For example, by forecasting future population growth and traffic demand, the layout of residential, commercial, and industrial areas can be reasonably planned to ensure smooth transportation connections between regions and improve the overall operational efficiency of the city.

(2) *Sustainable Urban Development*: Traffic demand forecasting helps promote sustainable urban development. By predicting future traffic demand, reasonable transportation policies can be formulated to reduce traffic congestion and environmental pollution. For example, by promoting public transportation, optimizing road design, and encouraging green travel, vehicle emissions can be reduced, air quality improved, and green urban development promoted.

**3. Improving Residents' Quality of Life**

Traffic demand forecasting is directly related to residents' travel experience and quality of life. Through accurate traffic demand forecasting, transportation systems can be optimized, travel efficiency improved, and residents' travel time and costs reduced. As shown in Figure 12.1, these are the effects brought by accurate traffic demand forecasting:

(1) *Improve Travel Efficiency*: Through traffic demand forecasting, traffic signal timing, road layout, and public transportation systems can be optimized to reduce traffic congestion and improve travel efficiency. Residents can enjoy more convenient travel services, spend less time on the road, and enhance their quality of life.

(2) *Reduce Travel Costs*: Traffic demand forecasting helps improve the service level of public transportation systems, encouraging more residents to choose public transit and reducing the use of private cars. This can lower residents' travel costs and ease the financial burden on households. For example, by optimizing bus routes and schedules, improving bus punctuality and comfort, more residents can be attracted to use buses, thereby reducing expenses on fuel and parking fees for private cars.

(3) *Enhance Travel Safety*: Traffic demand forecasting can help identify areas with frequent traffic accidents and high-risk road sections. Through reasonable traffic planning and management measures, the occurrence of traffic accidents can be reduced, improving residents' travel safety. For instance, by predicting and analyzing traffic flow, traffic enforcement can

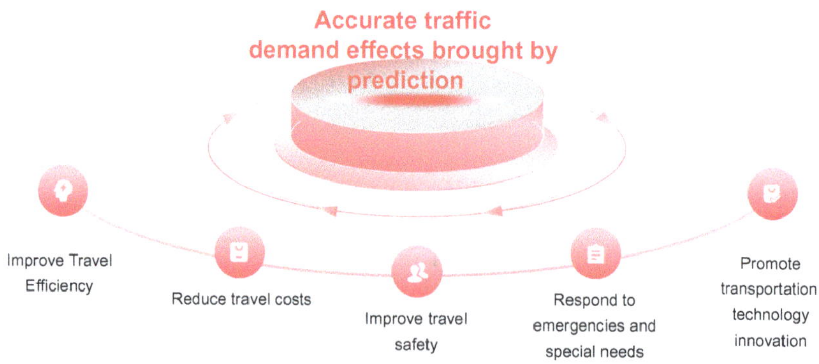


**Figure 12.1.** Effects of accurate traffic demand forecasting.

be increased during peak hours to ensure road safety; additional traffic signs and signals can be installed in high-risk areas to raise drivers' attention and reduce accidents.

(4) *Respond to Emergencies and Special Needs*: Traffic demand forecasting also plays an important role in responding to emergencies and special needs. By predicting future traffic demand, emergency plans can be formulated to ensure that the traffic system can respond quickly during unexpected events, maintaining the continuity and safety of urban operations.

(5) *Promote Transportation Technology Innovation*: Traffic demand forecasting is a key driver for promoting innovation in transportation technology. By continuously improving and optimizing traffic demand forecasting methods, the development of traffic management technologies can be advanced, enhancing the intelligence level of transportation systems.

Traffic demand forecasting relies heavily on large amounts of data support. Through big data analysis and artificial intelligence algorithms, the accuracy and reliability of forecasts can be improved. At the same time, traffic demand forecasting is also an important component of intelligent transportation systems. Accurate traffic demand forecasting enables the intelligent control of traffic signals, vehicle dispatching, and traffic management.

In summary, traffic demand forecasting plays an important role in traffic planning and management. It can improve the operational efficiency of traffic systems, support urban planning and development, and enhance residents' quality of life. Therefore, accurate traffic demand forecasting is a key link in achieving modern urban traffic management and planning, and BCI technology can also play a prominent role in improving the effectiveness of traffic demand forecasting.

### 12.1.2 *The Unique Advantages of Brain–Computer Interface Technology in Traffic Demand Data Collection*

The greatest feature of BCI technology is the collection of human brain signals, which is completely different from traditional data collection methods. In the era of big data, the main subjects of modern data collection are often objective data or historical data, while direct collection of human subjective, real-time data is very rare. BCI technology precisely fills this gap. By directly collecting the target's electroencephalogram (EEG) signals, it can reflect their cognitive state and behavioral patterns in real time and accurately, providing more comprehensive and dynamic data support for traffic demand forecasting. Next, we take a detailed look at the unique roles BCI technology can play in enhancing the effectiveness of traffic demand forecasting.

**1. Real-time Data Collection**

Real-time data collection is an important part of traffic demand forecasting. Traditional data collection methods, such as traffic flow monitoring and surveys, often suffer from data lag and limited coverage. BCI technology, by real-time monitoring of drivers' and passengers' EEG signals, can provide more timely and accurate traffic demand data.

BCI technology uses EEG signals to obtain real-time data on human brain activity. EEG sensors worn by drivers and passengers can monitor their cognitive states and emotional changes in real time. These sensors transmit the collected EEG signals wirelessly to the traffic management system for analysis and processing.

By real-time monitoring of EEG signals, BCI technology can capture these changes in traffic participants in real time and transmit the data to the traffic management center for analysis. Traffic management authorities can use this real-time data to promptly adjust traffic signal timing, optimize traffic flow, reduce traffic accidents, and improve road throughput efficiency.

**2. Precise Data Analysis**

Precise data analysis is key to improving the accuracy of traffic demand forecasting. BCI technology, by combining big data analytics and artificial intelligence algorithms, can deeply analyze the collected EEG signals, extract useful information, and provide a scientific basis for traffic demand forecasting.

The application of BCI technology in traffic demand data collection generates a large amount of EEG signal data. Through big data analysis techniques, key factors affecting traffic demand can be extracted from this data. For example, EEG data from different time periods and road sections can identify patterns of traffic flow changes and characteristics of peak periods, thereby providing precise data support for traffic demand forecasting.

Artificial intelligence algorithms play an important role in BCI data analysis. The combination of the two can classify, cluster, and predict large amounts of EEG signal data. In current research, many scholars use neural network algorithms to classify EEG signals into different cognitive and emotional states, thereby understanding the behavior patterns of traffic participants in various traffic environments and predicting future changes in traffic demand trends, providing a scientific basis for traffic planning and management.

However, the precise data collection and analysis of BCI technology require the integration of multiple disciplines, such as neuroscience, computer science, and traffic engineering. Only with the support of various technologies can the depth and breadth of data analysis be enhanced, thereby developing more intelligent and scientific traffic demand prediction models that improve the accuracy and reliability of forecasts.

In summary, BCI technology significantly improves the efficiency and accuracy of traffic demand data collection through real-time data acquisition and precise data analysis. This brings many benefits to traffic development, enhancing the quality of life for the public and contributing to the goal of building a strong transportation nation.

## 12.2 The Impact of Brain–Computer Interface on Traffic Network Design

Understanding traffic demand is the foundation for traffic network design, and the effectiveness of traffic network design directly determines the operational efficiency of a city and the quality of life of its residents. Currently, with China's economic development continuously improving, urban development levels rising, and city scales expanding, traffic demand is increasing and becoming increasingly complex. To meet the needs of modern, technological, and intelligent urban development, traffic network design methods must be continuously upgraded. Moreover, if we rely too heavily on historical data and static model design methods, it will be difficult to meet urban development demands. When facing rapidly changing traffic flows and constantly evolving travel demands, the humanization and intelligence level of the traffic network will also decline.

Above, we analyzed the application of BCI technology in traffic demand forecasting, demonstrating its great potential in improving prediction accuracy and dynamic adjustment capabilities. On this basis, we can also find that based on precise traffic demand data, BCI technology can further promote the optimization of traffic network design. Next, let us analyze the specific applications of BCI technology in traffic network design and understand effective methods for optimizing traffic network design.

## 12.2.1 *Comprehensive Considerations in Traffic Network Design*

Currently, all achievements in the development of China's comprehensive three-dimensional traffic network have been realized through advanced design planning and multiple rounds of optimization and improvement. Therefore, traffic network design is the first step in taking concrete development actions after understanding traffic construction demands, and this step holds an important position in traffic development.

Traffic network design is not only the professional work of traffic designers but also relates to various developmental considerations of cities and people's livelihoods. In the following, we explore in detail from different perspectives the relationships between traffic network design and the development of cities, regions, and even society.

**1. Design Considerations of Traffic Network Planners**

As the main planners of intelligent transportation, traffic network planners' primary task is to scientifically plan and reasonably layout the traffic network to ensure the efficient operation of the traffic system. Through detailed traffic demand analysis and data forecasting, planners can develop traffic network layout schemes that meet urban development needs. This includes road layout and classification, traffic signal system settings, and public transit route planning. Scientific planning and layout can effectively divert traffic flow, reduce congestion, and improve road capacity.

Second, another important goal of traffic network design is to improve the safety of the traffic system. Therefore, traffic network planners also need to reduce traffic accidents through reasonable road design and traffic signal settings.

At the same time, smart traffic network design needs to consider environmental protection and sustainable development. Through reasonable traffic planning, the development of green transportation modes can be promoted, motor vehicle usage reduced, and traffic pollution lowered. This is also a main consideration for traffic network planners.

**2. Design Considerations for City Administrators**

Traffic network design is an important support for urban economic development. An efficient traffic network can promote the smooth flow of logistics and people, enhancing the city's economic vitality. City administrators need to attract more enterprises and investments and promote rapid urban economic development through scientific traffic network design. For example, planning convenient traffic networks connecting commercial, industrial, and residential areas can facilitate various economic activities and improve the city's overall competitiveness.

Additionally, reasonable traffic network design helps improve urban governance capabilities. Through intelligent traffic systems and real-time traffic management, city administrators can better monitor and dispatch traffic flow, respond promptly to traffic emergencies, and enhance the efficiency of traffic management and emergency response.

Finally, city administrators also need to focus on urban livability. Through scientific traffic network design, they can improve residents' travel experience and enhance the quality of urban life. Convenient traffic networks can reduce residents' travel time and transportation costs, improving travel comfort and safety.

**3. Design Considerations for Urban Residents**

For urban residents, the direct impact of traffic network design is the convenience, efficiency, and safety of travel. A reasonable traffic network layout can greatly enhance residents' life rhythm and experience.

Additionally, traffic network design also determines residents' travel costs. Optimizing public transportation systems and non-motorized vehicle lanes, encouraging residents to choose low-cost travel modes such as buses, subways, and bicycles, and reducing private car usage are all main methods to lower urban residents' travel costs.

**4. Design Considerations for Economic Development**

Efficient transportation network design is of great significance to the development of the logistics industry. A convenient transportation network can enhance the attractiveness and commercial vitality of business districts. Through scientific transportation network design, the travel of residents and tourists can be facilitated, promoting the flow of people and consumption in business districts. For example, by setting up convenient bus routes and subway stations to connect business districts and residential areas, more consumers can be attracted to shop and entertain, thereby improving the economic benefits of business districts.

At the same time, transportation network design is an important infrastructure supporting urban industrial development. Through scientific transportation network planning, the construction and development of industrial parks and economic development zones can be promoted, enhancing the industrial competitiveness of cities.

**5. Design Considerations for Environmental Protection**

In recent years, China has consistently advocated for green transportation development, and the integration of green concepts is a major focus in current transportation network design.

Through reasonable transportation network design, the development of green travel modes can be promoted, improving the sustainable development level of cities. For example, setting up more non-motorized vehicle lanes and pedestrian paths within cities encourages residents to choose green travel modes such as cycling and walking, reducing the use of motor vehicles, which can effectively alleviate urban traffic pressure and environmental pollution.

Second, transportation network design also needs to consider ecological environment protection. Through reasonable planning and design, damage to the natural environment can be minimized. For example, when planning roads and transportation facilities, efforts should be made to avoid nature reserves and ecologically sensitive areas to reduce the impact on the ecological environment. This also helps protect the city's natural landscape and ecological resources, enhancing the city's environmental quality.

**6. Design Considerations for Social Equity**

Transportation network design needs to pay attention to social equity, ensuring that all residents can enjoy convenient transportation services. Through scientific transportation network design, social inclusion can be promoted, enhancing social harmony in cities. For example, the design of intelligent transportation networks will be equipped with barrier-free facilities and optimized public transportation system designs, providing convenient travel services for vulnerable groups such as the elderly, disabled, and children, improving their travel convenience and safety.

It can be seen that transportation network design is not a simple task for transportation departments; it relates to all aspects of the public, cities, and social development. Therefore, to ensure the effectiveness of transportation network design and to advance the goal of building a strong transportation nation, leveraging modern technological capabilities to optimize transportation network design is a major focus for contemporary transportation professionals.

### 12.2.2 *Current Status of Transportation Network Design in China*

Over the past few decades, China has made significant achievements in transportation network design. Major cities across the country have actively promoted the construction of transportation infrastructure and continuously optimized transportation network design, laying a solid foundation for improving the operational efficiency of urban transportation systems and enhancing residents' travel experience.

We can see that in recent years, across the country, scientific planning and reasonable layout in transportation network design have significantly improved road capacity and connectivity. For example, China's "Eight Vertical and Eight Horizontal" high-speed railway network has enhanced the quality of life for people nationwide and greatly strengthened the connection between residents in remote areas and developed cities. At the same time, it has had a profound impact on China's economic development. It has promoted the flow of goods and people across regions, reduced regional logistics costs, and improved production efficiency, greatly benefiting the development of China's agriculture and manufacturing industries. As of the end of November 2023, the main corridors of China's "Eight Vertical and Eight Horizontal" high-speed rail network have accelerated construction, with 36,100 kilometers completed and put into operation, accounting for about 80%, providing strong support for promoting coordinated regional development.

However, despite the many achievements China has made in transportation network design, with the continuous expansion of urban scale and the increasing demand for transportation, there are still areas and aspects of the existing transportation network design that need improvement. This is also an important direction for enhancing the effectiveness of China's transportation network design.

1. **Interdepartmental Coordination Needs to Be Strengthened:** Transportation network design is not solely the responsibility of the transportation department; it requires collaboration among multiple departments. Urban planning, environmental protection, public security, traffic management, and other departments need to work closely together to jointly formulate and implement transportation network design plans. Strengthening interdepartmental coordination helps comprehensively consider various factors and develop more scientific and reasonable transportation network designs. Currently, China's transportation network design still needs improvement in this area.

2. **Public Participation and Transparency Need to Be Improved:** The public are the direct users of the transportation network, and their opinions and needs are crucial to transportation network design. Enhancing public participation and transparency can improve the scientific and rational nature of transportation network design, as well as increase public understanding and support for transportation policies. Currently, China's transportation network design is still mainly led by transportation planning and design departments. Although transportation demand surveys have been conducted, their depth is insufficient. This is an area where China's transportation network design needs strengthening.

3. **The Forward-Looking Nature of Transportation Network Design Can Be Further Enhanced:** Transportation network design needs to be forward-looking and capable of adapting to future development demands. With the continuous advancement of urbanization, transportation demand will continue to grow, and transportation network design must have sufficient foresight to meet future challenges.

4. **The Intelligence and Innovation of Transportation Networks Can Be Further Improved:** The application of intelligent technology and innovative techniques is of great significance in transportation network design. For example, by introducing advanced intelligent transportation systems, the Internet of Things, big data analytics, and other technological means, the scientific and precise nature of transportation network design can be enhanced. Although China has also achieved remarkable results in smart transportation development, there is still a certain gap compared to

some developed countries. Therefore, this is also a key focus for improving China's transportation network design.

**5. The Green Attribute can Continue to be Strengthened:** Intelligent transportation network design must highly emphasize ecological environment protection and sustainable development. Through the construction of green transportation infrastructure and environmentally friendly design, the negative impact of transportation on the environment can be reduced, promoting the sustainable development of cities. To comprehensively strengthen the green attributes of China's transportation network, considering ecological environment protection and sustainable development is naturally a necessary task.

**6. Integration of Multimodal Transportation Systems Needs Improvement:** Modern urban transportation network design needs to promote the integration of multimodal transportation systems, enhancing the connection and transfer efficiency between different modes of transportation. How to achieve seamless connections among buses, subways, bicycles, walking, and other transportation modes through scientific planning and reasonable layout, thereby improving the overall operational efficiency of the transportation system, is also a current issue we need to address in transportation planning and design.

**7. International Exchange and Cooperation in Transportation Network Design Needs Enhancement:** One of the goals of China's transportation development is to reach an internationally advanced level, or even a leading position. This precisely requires strengthening international exchange and cooperation, learning from and drawing on advanced international transportation network design experiences to improve the current level and capability of China's transportation network design. Therefore, how to quickly understand and master cutting-edge international transportation network design concepts and technologies, and promote innovation and development in China's transportation network design, is also a problem we need to improve.

Overall, although China has made significant achievements in transportation network design, there are still many directions and spaces for improvement. Addressing and solving these issues can provide substantial help in enhancing the final development effect of China's transportation network from the design perspective.

### 12.2.3 *Brain–Computer Interface Technology Optimizes Transportation Network Design Effect*

To enhance the effectiveness of transportation network design, transportation departments and professionals worldwide are considering leveraging modern technology to strengthen the intelligence of transportation network design, striving to comprehensively improve the design and construction outcomes. In this process, the application of digital twin technology takes precedence. This technology realizes a real connection between the physical world and the virtual world. The use of digital models makes transportation network design more visual, and various tests and simulations in the digital world highlight the practicality and greenness of transportation network design. However, regarding human-centered features, many transportation technology experts have expressed that utilizing BCI technology can achieve better results and improve the overall level of transportation network design.

Previously, we also analyzed the growth potential and development directions of China's transportation network design. Now, let's take a detailed look at how BCI technology can complement digital twin technology in the design field to enhance the level of transportation network design.

**1. Strengthen Cross-Departmental Collaborative Cooperation**

BCI technology can achieve cross-departmental data sharing and collaboration, providing real-time data support needed by multiple departments such as traffic planning, urban management, and environmental protection. By establishing a unified data platform, BCI technology can collect and transmit EEG data in real time for reference and use by various departments, improving the efficiency and effectiveness of collaborative efforts.

With the precise data provided by BCI technology, it is also possible to assist cross-departmental collaboration in formulating more scientific and comprehensive traffic network design plans. Real-time EEG data can reflect the emotions, attitudes, and behaviors of traffic participants; these data further illustrate their demands on the traffic network, helping departments make decisions that align with actual conditions and enhancing the scientific and rational nature of traffic network design.

**2. Enhancing Public Participation and Transparency**

BCI technology can improve public enthusiasm and transparency in traffic network design through real-time data collection and analysis. By establishing a public participation platform, citizens can provide real-time feedback on their travel experiences and needs. BCI technology can analyze this feedback data to help traffic management departments optimize traffic network design.

Additionally, BCI technology can provide a more scientific basis for participation by collecting the public's EEG data. Compared to traditional questionnaires and opinion surveys, BCI technology can obtain more objective and authentic data, helping traffic management departments more accurately understand public needs and preferences.

**3. Strengthening the Forward-Looking Nature of Traffic Network Design**

BCI technology can enhance the forward-looking nature of traffic network design through long-term data accumulation and analysis. By continuously collecting and analyzing EEG data, traffic management departments can predict future traffic demands and flow trends, formulating forward-looking traffic network design plans. Traffic planning and design departments can analyze long-term data to forecast future traffic hotspot areas and changes in traffic flow, enabling early planning and construction of corresponding road infrastructure to avoid future traffic bottlenecks.

Moreover, BCI technology can serve as a forward-looking planning tool to assist urban planners in developing traffic network designs that adapt to future needs. Through real-time data and long-term forecasting, it can provide forward-looking planning recommendations to ensure that traffic network designs can meet the long-term demands of urban development.

**4. Integrating Intelligent Technology and Innovative Techniques**

BCI technology can be integrated with intelligent transportation systems to enhance the intelligence level of traffic network design. By combining digital twins, big data analytics, artificial intelligence, and Internet of Things technologies, BCI technology can monitor and analyze traffic flow and participant status in real time, optimizing traffic signal timing and road layouts. On this basis, traffic planning and design departments can innovate traffic management solutions, improving the scientific and effective nature of traffic network design.

**5. Protect the Ecological Environment and Promote Sustainable Development**

Through real-time data analysis, BCI technology can identify traffic participants' attitudes and needs toward the traffic environment, and then optimize road layouts and traffic flows based on actual conditions, reducing the environmental impact of traffic and promoting the development of green transportation methods. For example, by analyzing EEG data of public transport passengers, transportation planning and design departments can understand which public transit routes passengers are more satisfied with and which traffic environments cause resistance or aversion. Subsequently, through comparative analysis, they can design bus routes and construction plans that better meet passenger needs while optimizing the urban environment.

**6. Promoting the Integration of Multimodal Transportation Systems**

BCI technology can promote seamless connections among multimodal transportation systems, enhancing the coordination and transfer efficiency between different transportation modes. By collecting and analyzing real-time data from traffic participants, BCI technology can deeply identify bottlenecks and gaps between different transportation modes and optimize and supplement these areas accordingly. This ensures smooth connections and efficient transfers among various transportation methods.

**7. Strengthening International Exchange and Cooperation in Transportation Network Design**

Currently, various international seminars and exchanges on the development of BCI technology have been conducted. These transportation platforms can strengthen international exchange and cooperation in this technology, promoting its deep application across various fields. For example, the Tianqiao Brain Science Institute (TCCI) announced in July 2024 that it will jointly host the Brain–Computer Interface Society & Chen Institute Joint Brain–Computer Interface Meeting with the Brain–Computer Interface Society on December 6–7, 2024, in Shanghai, China. This is the first time in 25 years since the founding of the world's top academic conference in the BCI field that it will be held in Asia.

The Brain–Computer Interface Society is the most influential academic organization in the global BCI field. Its flagship event, the International Brain–Computer Interface Meeting, first held in New York, USA, in 1999, has become a benchmark academic conference in this field. The 10th meeting in 2023 was held in Brussels, Belgium, attracting nearly 500 scientists from 237 laboratories across 39 countries and regions worldwide.

Through such exchanges, we can learn from advanced international experiences in transportation network design, improve the level and capability of transportation network design in China, and promote innovation and development in China's transportation network design.

Additionally, by cooperating with leading international transportation research institutions and enterprises, we can also introduce advanced international technologies and equipment, conduct joint research and development, and promote the internationalization and modernization of China's transportation network design.

In summary, BCI technology can significantly optimize the effectiveness of transportation network design, providing strong support and assurance for the development of modern urban transportation. This is very important for the scientific, intelligent, and efficient upgrading of China's transportation network design, as well as for the sustainable and human-centered development of urban transportation.

## 12.3 The Impact of Brain–Computer Interfaces on Transportation Policy Making

BCI technology has brought disruptive changes to urban traffic planning. By capturing and analyzing the EEG signals of traffic participants in real time, BCI technology provides a scientific basis for traffic planning decisions, optimizes the design and management of urban traffic networks, promotes the modernization of traffic systems, and has a profound impact on traffic policy formulation. In the following, we analyze how this technology enhances the rationality and effectiveness of traffic policy making.

### 12.3.1 *Traffic Policies Promote Traffic Development*

In recent years, the Ministry of Transport of China has issued a series of key policies aimed at promoting traffic development, optimizing traffic systems, and improving traffic management levels. These policies not only provide scientific guidance for the construction of intelligent transportation systems but also play important roles in promoting economic development, improving residents' lives, and protecting the ecological environment. Among them, the "Outline for Building a Strong Transportation Country," the "National Comprehensive Three-Dimensional Transportation Network Planning Outline," and the "14th Five-Year Plan for the Development of a Modern Comprehensive Transportation System" are the most critical. They serve as decisive policy bases for maintaining high-quality and rapid development of China's transportation and play a core leading role in the country's traffic development.

The "Outline for Building a Strong Transportation Country," formulated in 2019, is a strategic document aimed at achieving the goal of a strong transportation nation. The outline clarifies the overall goals, development directions, and key tasks for building a strong transportation country, aiming to promote high-quality development of China's transportation industry and build a world-class modern transportation system. The policy explicitly states that by 2035, a strong transportation country will be basically established. A modern comprehensive transportation system will be basically formed, public satisfaction will significantly improve, and the capacity to support national modernization will be greatly enhanced; it will have a developed express network, a complete trunk network, and an extensive basic network, with coordinated development of urban and rural regional transportation reaching new heights; the "National 123 Travel Traffic Circle" (1-hour commute in metropolitan areas, 2-hour access in city clusters, and 3-hour coverage of major cities nationwide) and the "Global 123 Fast Freight Logistics Circle" (1-day delivery domestically, 2-day delivery to neighboring countries, and 3-day delivery to major global cities) will be basically formed, enabling convenient and smooth passenger intermodal transport and efficient, economical multimodal freight transport; the development levels of intelligent, safe, green, and shared transportation will be significantly improved, urban traffic congestion will be basically alleviated, and barrier-free travel service systems will be basically perfected; the transportation science and technology innovation system will be basically established, key transportation equipment will be advanced and safe, talent teams will be excellent, and the market environment will be favorable; the modernization of the transportation governance system and governance capabilities will be basically realized; and the international competitiveness and influence of transportation will be significantly enhanced.

This policy emphasizes the importance of enhancing transportation science and technology innovation capabilities. By increasing investment in technology, promoting technological innovation and industrial upgrading in the transportation sector, it aims to improve the independent innovation capacity and international competitiveness of transportation technology. For example, the policy proposes accelerating the development of intelligent transportation, green transportation, and safe transportation technologies, and promoting the transformation and application of scientific and technological achievements in the transportation field. This not only enhances transportation technology innovation capabilities but also drives the high-quality development of the transportation industry.

The "Outline for Building a Strong Transportation Country" also sets the goal of constructing an efficient and convenient transportation network. By optimizing the layout of the transportation network and improving the connectivity and service levels of transportation infrastructure, it aims to build a modern, efficient, and convenient transportation network.

Additionally, the policy emphasizes the importance of promoting international cooperation in transportation. By strengthening international exchanges and cooperation, it aims to enhance China's international influence and competitiveness in transportation and promote the internationalization of the transportation industry. The policy points out the need to actively participate in the "Belt and Road" initiative, strengthen transportation cooperation with countries along the route, and promote the construction and operation of international transportation projects. This not only enhances China's international competitiveness in transportation but also facilitates international transportation cooperation and exchanges.

The "National Comprehensive Three-Dimensional Transportation Network Planning Outline," formulated in 2021, is a strategic planning document aimed at constructing a comprehensive three-dimensional transportation network. The outline clarifies the overall layout, development goals, and key tasks of the comprehensive three-dimensional transportation network.

This policy first summarizes and analyzes the development of China's comprehensive three-dimensional transportation network. It clearly states that since the reform and opening up, especially since the 18th National Congress of the Communist Party of China, under the strong leadership of the Party Central Committee with Comrade Xi Jinping at its core, China's transportation development has achieved remarkable accomplishments. The infrastructure network has basically taken shape, and the comprehensive transportation system has been continuously improved; transportation service capacity and levels have significantly increased, and the public's sense of gain has been notably enhanced; technological innovation achievements are remarkable, and the technical levels of facility construction and transportation equipment have greatly improved; modernization of transportation construction has accelerated, and the levels of safe, intelligent, and green development have continuously risen; the openness of transportation to the outside world has expanded steadily, and the pace of going global has accelerated. Transportation development has effectively promoted the development and protection of territorial space, coordinated urban–rural and regional development, and optimized the layout of productive forces, playing a fundamental, pioneering, strategic, and service role in economic and social development, providing strong support for the decisive victory in building a moderately prosperous society in all respects.

At the same time, there are still some shortcomings in China's transportation development, with prominent issues of imbalance and insufficiency. The layout of the comprehensive transportation network still needs improvement, the structure requires optimization, and connectivity and network resilience need to be strengthened; the integrated coordination of comprehensive transportation urgently needs enhancement, the level of resource-intensive utilization needs to be raised, the collaborative integration of transportation with related industries needs to be deepened, and the support capacity of the entire industry chain still needs improvement; the quality, efficiency, and service levels of comprehensive transportation development are not high, the modern logistics system needs to be improved, and the capabilities for technological innovation, safety, intelligence, and green development need further enhancement; reform tasks in key areas and critical links of transportation remain arduous.

On this basis, the "National Comprehensive Three-dimensional Transportation Network Planning Outline" puts forward specific requirements for optimizing the layout of the comprehensive transportation network. By reasonably planning and arranging various transportation modes such as highways, railways, aviation, and water transport, an interconnected comprehensive transportation network is constructed. The policy explicitly proposes the construction of a national high-speed railway network, a national expressway network, a nationwide airport cluster, and port clusters to enhance the overall connectivity and coordination of the transportation network. This optimizes the allocation of transportation resources, improves transportation efficiency, and promotes regional coordinated development.

In this policy, the Chinese government emphasizes the coordinated development of the transportation system. By promoting the integrated development of multiple transportation modes, it aims to enhance comprehensive transportation capacity and service levels. For example, the policy proposes strengthening the connection between highways and railways, aviation and railways, ports and railways, and other transportation modes, developing multi-modal transport systems to improve the efficiency and service quality of comprehensive transportation. These measures enhance the overall coordination and operational efficiency of the transportation system.

The "14th Five-Year Plan for the Development of a Modern Comprehensive Transportation System" was released in December 2021 as a guiding document for transportation development during the 14th Five-Year Plan period. This plan clarifies the development direction, goals, and tasks of the transportation system over the next five years, aiming to build a safer, more efficient, greener, and smarter modern comprehensive transportation system.

This policy sets specific goals and tasks for accelerating the construction of transportation infrastructure. It clearly stipulates that by 2025, comprehensive transportation will basically achieve integrated and coordinated development, with substantial breakthroughs in intelligence and greenness. Comprehensive capacity, service quality, operational efficiency, and overall benefits will be significantly improved, advancing transportation development to a world-class level.

The infrastructure network is more complete. The utilization rate of the main framework capacity of the national comprehensive three-dimensional transportation network has significantly increased. Using the "Eight Vertical and Eight Horizontal" high-speed railway main corridors as the backbone, connected by regional high-speed railway lines and supplemented by intercity railways that partially serve trunk line functions, the high-speed railway network with speeds of 250 km/h and above covers over 95% of cities with populations exceeding 500,000. Bottleneck sections of conventional railways have been basically eliminated. The main lines of the national expressway network, composed of 7 capital radial lines, 11 north-south vertical lines, 18 east-west horizontal lines, as well as regional ring lines, parallel lines, and connecting lines, are basically connected. The quality of ordinary roads has further improved. A well-planned and fully functional modern airport system has basically formed. The specialization and modernization level of ports and wharves have significantly improved, and important progress has been made in the construction of a high-grade inland waterway network. The efficiency of transfers and transshipments at comprehensive transportation hubs has further increased. Integrated transportation networks for key city clusters and one-hour commuting networks for metropolitan areas are accelerating, and border national highways are basically connected.

Transportation services are more efficient. The quality of transportation services has steadily improved, with passenger "one-stop" and freight "single-order" services becoming more widespread. Customized, personalized, and specialized transportation service products have become richer. Urban traffic congestion and parking difficulties continue to ease, transportation services in rural and border areas are better guaranteed, and express delivery services have achieved full coverage in eligible administrative villages. The global international transportation service network is more complete, and the quality of the China–Europe Railway Express has steadily improved.

Technological equipment is more advanced. Technologies such as fifth-generation mobile communication (5G), the Internet of Things, big data, cloud computing, and artificial intelligence are deeply integrated with transportation. Significant progress has been made in the construction of new infrastructure in the transportation sector, with a marked increase in the digitalization rate of transportation infrastructure. Substantial breakthroughs have been achieved in data openness, sharing, and platform integration and optimization. The promotion and application of domestically developed advanced technological equipment have accelerated, achieving comprehensive coverage of key transportation areas by the BeiDou system, and the standardization rate of transportation equipment has greatly improved.

Safety guarantees are more reliable. Transportation facilities are durable and reliable, operations are safe and controllable, and preventive measures are in place. The integrity rate of safety facilities continues to improve. Cross-departmental and cross-sector safety risk prevention and control systems, as well as emergency rescue systems, have been further improved, with a further reduction in the incidence of major and particularly serious accidents. The safety of transportation on main corridors and the transportation of goods such as grain, energy, and ores is better ensured, and the security capability of international logistics supply chains continues to improve.

The development model is more sustainable. Green production and lifestyle practices in the transportation sector are gradually taking shape. The proportion of bulk and medium-to-long-distance freight transported by rail and waterway steadily increases. The share of green travel has significantly risen, clean and low-carbon transportation tools are widely used, energy consumption per unit turnover has significantly decreased, the proportion of green construction in transportation infrastructure has markedly increased, resource utilization efficiency continues to improve, and carbon emission intensity steadily declines.

Governance capabilities are more complete. The integrated development of various transportation modes, laws, regulations, and standards in areas such as transportation infrastructure investment, financing, management, operation, and maintenance has been further improved. The degree of integrated development of comprehensive transportation continues to increase, market-oriented reforms deepen continuously, diversified investment and financing systems become more sound, and a new credit-based regulatory mechanism is accelerating its formation.

Looking ahead to 2035, a convenient, smooth, economical, efficient, safe, reliable, green, intensive, intelligent, and advanced modern high-quality national comprehensive three-dimensional transportation network will be basically established. The "National 123 Travel Circle" (1-hour commuting in metropolitan areas, 2-hour access in city clusters, and 3-hour coverage of major cities nationwide) and the "Global 123 Express Freight Circle" (1-day domestic express delivery, 2-day delivery to neighboring countries, and 3-day delivery to major global cities) will be basically formed, marking the basic establishment of a strong transportation nation.

The "14th Five-Year Plan" emphasizes the development of intelligent transportation systems. By introducing advanced technologies such as big data, the Internet of Things, and artificial intelligence, it aims to enhance the intelligence level of traffic management. The policy also clearly defines the direction for promoting green transportation development. Through measures such as promoting new energy vehicles, optimizing the energy structure of transportation, and reducing traffic pollution emissions, it facilitates the green and low-carbon transformation of the transportation system. Overall, this policy clarifies the clear directions and measures for building a modern comprehensive transportation system, providing strategic support for the comprehensive construction of a socialist modern country.

Reviewing the development process of China's transportation over the past few years, a series of key policies issued by the Ministry of Transport have provided scientific guidance for the progress of transportation in China. These policies have also played an important role in promoting economic development, improving residents' lives, and protecting the ecological environment. The implementation of these policies has promoted the high-quality development of China's transportation industry, advanced the construction of a modern transportation system, and laid a solid foundation for achieving the goal of building a strong transportation country.

### 12.3.2 *Key Factors in Transportation Policy Formulation*

Through the analysis of key policies such as the "14th Five-Year Plan for the Development of a Modern Comprehensive Transportation System," the "National Comprehensive Three-Dimensional Transportation Network Planning Outline," and the "Outline for Building a Strong Transportation Country," we can deeply understand the importance and profound significance of transportation policy in national development and transportation progress.

However, formulating truly promotive transportation policies is not a simple task. To ensure the scientificity, rationality, and effectiveness of policies, multiple key factors must be considered and comprehensively evaluated. This includes reliance on accurate data and information, support for technological innovation, comprehensive assessment of socio-economic factors, and adherence to laws, regulations, and standards. Only by deeply understanding and grasping these key factors can feasible and long-term impactful transportation policies be formulated.

Next, we analyze the key factors in transportation policy formulation in detail, so as to better understand the policy-making process and clarify how transportation policies can effectively promote the optimization and improvement of transportation systems.

**1. National Economic Development Strategy**

The national economic development strategy is the primary consideration in transportation policy formulation. As an important tool to promote economic development, transportation policy must be consistent with the overall national economic strategy. For example, when implementing the national regional coordinated development strategy, transportation policy may prioritize strengthening transportation infrastructure construction between regions to promote regional economic interconnection. This strategy not only improves traffic conditions but also promotes the integrated development of regional economies and reduces economic disparities between regions.

Moreover, different stages of national economic development will impose different requirements on transportation policies. During periods of rapid economic growth, transportation policies may focus more on infrastructure construction and capacity enhancement to meet the needs of economic development. In relatively stable economic stages, transportation policies may emphasize improving transportation efficiency and service quality. By closely aligning with the national economic development strategy, transportation policies can more effectively support economic growth and promote social progress.

**2. Level of Economic Development and Stage of Transportation Industry Development**

Transportation policy formulation must fully consider the country's level of economic development and the development stage of the transportation industry. At different stages of economic development, transportation demand and the characteristics of the transportation industry's development vary. During periods of rapid economic growth, transportation demand is high, and there is a strong need for infrastructure construction. At this time, transportation policy should focus on large-scale investment and construction of transportation infrastructure to enhance transportation capacity and meet the needs of economic development. In contrast, during relatively stable economic development stages, transportation demand tends to stabilize, and the focus of the transportation industry may shift to improving transportation efficiency and service quality. At this stage, transportation policy should place greater emphasis on optimizing the transportation network and enhancing the level of transportation services and management efficiency.

**3. Geographic and Resource Conditions**

Geographic and resource conditions are important factors that cannot be ignored in the process of formulating transportation policies. The geographic environment and resource conditions vary greatly across different regions, which directly affect the site selection, construction, and operating costs of transportation infrastructure. For example, the construction of transportation infrastructure in mountainous and coastal areas needs to consider multiple factors such as terrain, geology, and climate, while resource-rich areas may be more inclined to develop transportation policies related to resource transportation.

In mountainous areas, the construction of transportation infrastructure must overcome the limitations of terrain and geological conditions, resulting in high costs and great difficulty. Therefore, transportation policies in these regions may focus on projects with high investment returns and consider adopting new technologies and materials to reduce construction and maintenance costs. Coastal areas, due to their flat terrain, have relatively easier transportation infrastructure construction but need to address the impacts of marine climate and tides. Transportation policies should emphasize the formulation and implementation of disaster prevention and mitigation measures.

**4. Social Factors**

Social factors include social values, culture, customs, moral concepts, etc., which directly influence the public's acceptance and willingness to participate in transportation policies. In an era with strong environmental awareness, the public may be more inclined to support green transportation policies. Governments need to consider the public's environmental protection demands when formulating transportation policies, promoting public transportation and low-carbon travel modes. In societies with strong traditional views, policy-makers need to consider how to promote transportation modernization while respecting traditional customs.

**5. Safety and Environmental Protection Requirements**

With social development and progress, safety and environmental protection requirements have become key factors influencing the formulation of transportation policies. Transportation policies need to fully consider traffic safety and environmental protection issues by formulating scientific and reasonable policy measures to reduce the incidence of traffic accidents and minimize environmental pollution caused by traffic emissions.

Regarding traffic safety, policy-makers need to improve traffic safety levels through legislation, standard setting, technology application, and other means. For example, by implementing strict traffic safety regulations, increasing penalties for traffic violations, promoting advanced traffic safety technologies and equipment, and enhancing the safety performance of roads and vehicles, the occurrence of traffic accidents can be reduced.

In terms of environmental protection, policy-makers need to formulate and implement a series of environmental policies and measures to reduce the negative impact of transportation on the environment. For example, promoting new energy vehicles to reduce the use of traditional fuel vehicles; optimizing transportation networks to reduce vehicle congestion and idle time; and encouraging the public to choose green travel modes such as public transportation and bicycles to reduce traffic emissions and environmental pollution.

**6. Technological Progress and Innovation**

Technological progress and innovation are important factors driving the formulation of transportation policies. With the development and application of technology, new modes of transportation and vehicles continuously emerge, providing new ideas and possibilities for transportation policy formulation.

Intelligent transportation systems improve the efficiency and quality of traffic management and services through the application of modern information technology. By collecting and analyzing real-time traffic information, optimizing traffic signal control, and reducing traffic congestion; by promoting intelligent public transportation systems to enhance the convenience and attractiveness of public transit and reduce private car usage; and by promoting vehicle-to-everything (V2X) technology to enable information sharing and coordination among vehicles, thereby improving traffic safety and driving efficiency.

**7. International Situation and External Environment**

Changes in the international situation and external environment also affect the formulation of transportation policies. For example, developments in international trade and changes in international logistics demand directly influence transportation policy formulation. At the same time, the formulation and changes of international transportation regulations also have a certain impact on China's transportation policy development.

In the context of globalization, transportation policies need to consider how to adapt to the development needs of international trade and logistics. This includes building and improving international logistics corridors to enhance logistics efficiency, reduce logistics costs, and strengthen the competitiveness of international trade; participating in the formulation of international transportation regulations to enhance China's voice and influence in international transportation affairs. Meanwhile, policy-makers need to pay attention to the latest trends and technologies in international transportation development, learn from advanced international experiences, and promote the modernization and internationalization of China's transportation.

In summary, the key factors influencing transportation policy formulation are very diverse. These factors intertwine and interact, jointly determining the direction and focus of transportation policy formulation. During the policy-making process, policy-makers need to comprehensively consider these factors and scientifically and reasonably formulate transportation policies to promote the sustainable development of transportation and support the comprehensive progress of the economy and society.

### 12.3.3 *The Impact of BCI Technology on Transportation Policy Formulation*

Having understood the core values and basis for transportation policy formulation, we can more clearly identify the main impacts of BCI technology on transportation policy development. Although BCI is only one technological force in the transportation field, this technology brings many innovations to transportation development, optimizing traffic management and providing more ideas for policy formulation. Next, let us explore how BCI technology can assist in the effective formulation of transportation policies.

**1. Assisting in the Formulation of Traffic Safety Policies:** BCI technology has great potential in enhancing traffic safety. By monitoring drivers' brainwaves in real time, BCI systems can detect drivers' fatigue status, attention levels, and emotional changes, thereby preventing potential traffic accidents. Policy-makers can promote intelligent driving assistance systems based on BCI technology, where the system automatically issues warnings or takes control of the vehicle when driver fatigue or inattention is detected. The application of such technology will significantly improve traffic safety levels and reduce traffic accidents caused by human factors.

**2. Optimization of Traffic Management Policy Formulation:** The application of BCI technology in traffic management can significantly optimize traffic flow and improve traffic efficiency. By integrating EEG signals with traffic management systems, traffic authorities can more accurately predict traffic congestion and flow changes, and adjust traffic signal timings and lane usage strategies in real time. For example, during peak hours, a BCI-based traffic management system can dynamically adjust signal durations and lane allocations by recognizing drivers' stress levels and traffic flow variations, thereby enhancing traffic smoothness. Policy-makers can promote the adoption of such intelligent traffic management systems to improve urban traffic management levels.

**3. Promote the Formulation of Barrier-Free Transportation Policies:** The application of BCI technology in barrier-free transportation can provide greater convenience for people with disabilities. By controlling vehicles or other transportation means through brainwaves, individuals with disabilities can achieve more independent travel. Policy-makers can formulate relevant policies to promote and popularize such barrier-free transportation tools, thereby improving the quality of life and social participation of people with disabilities.

**4. Support the Development of Intelligent Transportation Systems:** The application of BCI technology in intelligent transportation systems can further enhance the intelligence level of traffic systems. By integrating BCI technology with the Internet of Vehicles, big data analytics, artificial intelligence, and other technologies, transportation systems can operate and be managed more efficiently. Policy-makers can promote the development of such intelligent transportation systems to improve the overall intelligence and service level of the transportation network.

**5. Enhance Public Participation in Transportation Policy:** BCI technology can also enhance public participation in the transportation policy-making process. Through BCI technology, governments and traffic management departments can better understand the public's transportation needs and preferences. Governments can use BCI technology to conduct public opinion surveys, collecting genuine feedback and suggestions on transportation policies. This EEG-based survey method can reduce biases inherent in traditional questionnaires, obtaining more accurate and authentic public opinions, thereby improving the scientific basis and credibility of transportation policies.

**6. Promote Policies for Emerging Transportation Modes:** The application of BCI technology can also promote the development of emerging transportation modes, such as autonomous driving and shared mobility. Autonomous driving systems based on BCI technology can achieve higher levels of safety and intelligence, which requires relevant policies to plan and guide the development of these emerging transportation modes.

Currently, China's policies on autonomous vehicles are gradually being improved, which has promoted the development and application of autonomous driving technology. At the national level, regulatory bodies such as the State Council have issued a series of policies, including the "14th Five-Year National Informatization Plan" and the "14th Five-Year Digital Economy Development Plan," focusing on expanding autonomous driving application scenarios and providing guidance and specific plans for infrastructure construction, data security, and key development directions. At the local level, Beijing has issued the "Beijing Autonomous Vehicle Regulations (Draft for Comments)," which explicitly proposes reserving space for intelligent roadside infrastructure in new, renovated, and expanded roads, and supports the circulation and utilization of autonomous vehicle data. Additionally, Beijing has issued road testing licenses to 31 testing companies, with autonomous driving test mileage exceeding 28 million kilometers. Within the year, the city aims to achieve full coverage of intelligent roadside infrastructure across 600 square kilometers. In terms of specific policy measures, the Chinese government encourages qualified localities to conduct pilot operations and commercial services for autonomous vehicle sharing, shuttle connections, and intelligent parking, strengthens safety risk prevention and control, conducts pre-assessment studies on operational and management risks arising from autonomous driving technology applications, accelerates the creation of a favorable policy environment, improves management standards for autonomous vehicle road testing, and explores the establishment of a safety supervision system for the operation of autonomous commercial vehicles.

These policy measures jointly promote the development and application of autonomous driving technology, providing policy support for the functional optimization and innovation of intelligent connected vehicle products, and contributing to the rapid development of the intelligent connected vehicle industry and the construction of the industrial ecosystem.

In summary, brain–machine interface technology has a significant impact on the formulation of transportation policies. Policy-makers can fully leverage the characteristics and advantages of brain–machine interface technology to actively promote its application and development in the transportation field. Through scientific and reasonable policy planning and formulation, the safety, efficiency, and service level of transportation systems can be improved, thereby promoting the sustainable development of the transportation sector.

# Chapter 13

# Future Prospects of Brain–Machine Interface Technology's Impact on the Transportation Field

## 13.1 Development Trends of Brain–Machine Interface Technology

Since the concept was first proposed by American scientist Jacques Vidal in 1973, half a century has passed. Although brain–machine interface (BCI) technology was born relatively early, its development showed no significant progress for a long time. At the 2023 World Artificial Intelligence Conference's "Brain–Machine Intelligence and Digital Life" themed forum, Peng Lei, co-founder and CEO of BrainTiger Technology, stated that the lack of unification in BCI software, hardware, data, and algorithms is seriously restricting global collaboration in brain science. Current BCI technology is still in its infancy, and its future impact across various fields will be immeasurable.

As of 2024, BCI technology has become a global focus of scientific research and development. For example, Elon Musk's BCI company Neuralink completed the first human implantation of a BCI device in January 2024, with the recipient currently recovering well. This device, named "Telepathy," once implanted in the brain, allows control of phones and computers purely through thought, and through them, control of almost all devices. Even people who have lost limb function can operate various devices using this product. It is imaginable how profoundly this technology will change human development. As Musk said, super artificial intelligence is inevitable, and humanity has only one choice: to become AI. The intermediate goal of brain–machine interface technology is to enable human–AI integration and avoid being surpassed by AI. The rapid global development of AI actually relies on some institutions and organizations completing standardization work in the aforementioned fields.

In recent years, with the deepening research of global scientists on brain–machine interface technology, significant progress has been made. Current research mainly focuses on two directions: invasive and non-invasive brain–machine interfaces. Invasive brain–machine interfaces achieve high-precision neural signal acquisition and decoding by directly implanting electrode arrays into the cerebral cortex; non-invasive brain–machine interfaces collect signals on the scalp surface through methods such as EEG. Although their signal accuracy is lower, they offer higher safety and ease of use.

Currently, brain–machine interface technology has shown application potential in multiple fields. For example, in the medical field, brain–machine interface technology has been used to help paralyzed patients achieve limb control; in the entertainment field, it is gradually being applied to gaming and virtual reality. In the transportation field, brain–machine interface technology has also begun to play roles at various levels, such as traffic management, vehicle operation, traffic safety improvement, and transportation policy formulation, with its future potential being considerable.

Based on the application characteristics of brain–machine interface technology and current research achievements, we can see that breakthroughs are needed in the following areas to bring profound impacts across various fields.

**1. Improvement of Signal Acquisition and Processing Technology:** With the development of materials science and microelectronics technology, new sensor and electrode materials will greatly improve the quality and stability of signal acquisition. For example, flexible electronics technology will make electrodes fit the scalp better, reducing noise and interference. In addition, advanced signal processing algorithms and artificial intelligence technologies will further enhance the accuracy and efficiency of signal decoding, thereby achieving more reliable brain–computer interface (BCI) systems.

**2. Development of Non-invasive Technologies:** Although invasive technologies have advantages in signal acquisition accuracy, their surgical risks and costs are relatively high. In the future, non-invasive BCI technologies will become mainstream, especially for everyday transportation applications. Non-invasive technologies including EEG, near-infrared spectroscopy (NIRS), and ultrasound will achieve breakthroughs in signal acquisition accuracy and convenience, making BCI devices more suitable for controlling and monitoring daily transportation tools.

**3. Optimization of Real-time Performance and Response Speed:** The real-time performance and response speed of BCI systems are key factors affecting user experience. Future research will focus on reducing system latency and improving response speed, enabling users to interact with devices more naturally.

**4. Enhancement of Personalization and Adaptability:** Since individual differences exist in each person's brainwave signals, the future development of BCI systems will place greater emphasis on personalized and adaptive design. Through machine learning and adaptive algorithms, systems will automatically adjust parameters to meet the needs of different users, improving ease of use and user experience.

Overall, although BCI technology is not a recently invented technology, it has only recently undergone significant development. The slow progress in its early stages was due to the lack of support from other technologies. However, with the emergence and development of big data, AI, cloud computing, and other technologies, the advancement of BCI technology has received substantial support. With continued research and development, it is believed that this technology can better serve the progress and development of human society.

## 13.2 Application of Brain–Computer Interfaces in Transportation Vehicle Design

Imagine that future transportation vehicles can not only understand your commands but also read your intentions and anticipate your needs. This is not a plot from a science fiction novel but a real door opened by BCI technology. By decoding brain signals, BCI technology enables seamless interaction between humans and machines, completely overturning the traditional design concepts of transportation vehicles. This cutting-edge technology with unique application value will revolutionize the design and development of transportation vehicles and create unprecedented travel experiences. In the following, we analyze the application scenarios of this technology in future transportation vehicles.

**1. Intelligent Systems to Enhance Driving Safety**

In the design of transportation vehicles, driving safety has always been one of the primary considerations. BCI technology can effectively prevent safety hazards such as fatigue driving and inattention by monitoring the driver's EEG activity in real time.

BCI technology can embed EEG sensors into the driver's headgear or seat headrest to analyze the driver's attention and alertness status in real time. Once the system detects signs of fatigue or decreased attention, it will issue an alert and take corresponding measures, such as automatically adjusting the driver assistance system or suggesting a rest.

**2. Personalized Driving Experience**

BCI technology can also be used to create a personalized driving experience. Every driver has different driving habits and preferences, and traditional driving systems find it difficult to meet everyone's needs. Through BCI technology, vehicles can automatically adjust driving parameters based on the driver's EEG signals, such as seat position, rearview mirror angle, and air conditioning temperature. Additionally, the BCI system can record the driver's emotional changes and adjust music playback, interior lighting, and other settings according to the emotional state to enhance the driver's comfort and enjoyment. This personalized driving experience will make drivers feel that the vehicle better fits their needs, thereby increasing driving pleasure and comfort.

**3. Assistance and Enhancement of Autonomous Driving Technology**

In the development of autonomous driving technology, BCI technology also plays an important role. Although current autonomous driving systems have a high degree of autonomy, driver participation and decision-making are still required in some complex or emergency situations. BCI technology can enable seamless switching between human and vehicle control, quickly transmitting the driver's intentions and commands when the system detects that the driver needs to take over.

**4. Intelligent Assistance and Information Feedback**

In the future design of transportation vehicles, BCI technology will be widely applied in intelligent assistance systems and information feedback mechanisms. For example, through BCI technology, vehicles can obtain the driver's intentions and needs in real time, automatically adjusting navigation routes, speed, driving modes, and more. This intelligent assistance system can not only improve driving efficiency but also significantly enhance the driver's operational convenience and safety. Furthermore, BCI technology can be used in in-vehicle information feedback systems to understand the driver's needs and feedback through EEG signals in real time. When the driver needs to find a specific location or information, they only need to issue a mental command, and the system will immediately respond and display the relevant information. This intuitive interaction method will allow drivers to focus more on driving and reduce interference from operating in-vehicle devices.

**5. Design and Development of New Types of Transportation Vehicles**

The application of BCI technology is not limited to optimizing traditional transportation vehicles but will also give rise to new types of intelligent transportation vehicles. It is believed that in future development, smart cars, bicycles, motorcycles, and even intelligent flying vehicles will become widely popular, all integrated with BCI technology. These vehicles can be controlled through EEG signals to achieve precise direction control, speed adjustment, and automatic obstacle avoidance functions.

In addition, BCI technology can also be used to develop new types of shared transportation tools, attracting more users through personalized usage experiences, thereby promoting the development of the sharing economy.

The continuous advancement and expanding application scope of BCI technology will have an increasingly profound impact on vehicle design. In the future, BCI technology is expected to become an essential component of intelligent transportation vehicles, providing drivers with a safer, more comfortable, and smarter driving experience. As the technology matures and becomes widespread, BCI will drive innovation in vehicle design, creating more convenient, efficient, and environmentally friendly modes of transportation.

## 13.3 Integration of Brain–Computer Interface and Wearable Devices

Today, when we talk about the topic of future intelligent life, we can imagine various scenarios from current technological trends and even science fiction movies. However, most of these relaxed, convenient, and highly intelligent scenarios are related to wearable devices, and the detailed services provided by these wearable devices often require BCI technology to realize.

For example, a lightweight head-worn device can not only monitor your health status but also read your thoughts, coordinating various devices in real time to provide timely services. Such a science-fiction-like scenario is gradually becoming a reality. The integration of BCI technology with wearable devices will bring unprecedented intelligence and personalization to life, creating truly "understanding" service facilities for you. Next, let us imagine what surprises the combination of BCI technology and wearables could bring to the future transportation field.

### 1. Intelligent Head-Worn Devices and Driving Assistance

Intelligent head-worn devices, as one of the representatives of wearable devices, have been applied in many fields. For example, modern AI gaming glasses and helmets belong to intelligent head-worn devices, which allow users to experience unimaginable speed and excitement in the digital world.

In the process of integrating BCI technology, these devices can also become driving instruments for vehicles. Through EEG sensors, head-worn devices can monitor the driver's driving status to ensure driving safety. Additionally, they can control navigation, answer calls, play music, and more via brainwaves, achieving fully hands-free operation and enhancing the driving experience.

### 2. Health Monitoring and Driving Optimization

Modern wearable devices such as smartwatches and fitness bands already have basic health monitoring functions. Integrating BCI technology into these devices can further enhance their functions and application scope. For example, by monitoring the driver's heart rate, brainwaves, and galvanic skin response, BCI devices can assess the driver's stress level and emotional state in real time. When detecting that the driver is under high stress or experiencing significant emotional fluctuations, the system can suggest the driver take a break or perform relaxation exercises, and even automatically adjust the vehicle's air conditioning and music to help the driver relax and maintain optimal driving condition.

**3. Seamless Human–Machine Interaction and Personalized Customization**

Combining BCI technology with wearable devices will achieve an unprecedented human–machine interaction experience. By reading the user's EEG signals, transportation tools can perceive the user's needs and intentions in real time and make corresponding adjustments. This seamless interaction not only improves operational convenience but also greatly enhances the user experience. In addition, the personalized customization function based on the user's EEG data will enable each vehicle to provide tailor-made services for its owner.

**4. Remote Control**

The combination of BCI technology and wearable devices can also be applied to remote control and collaborative work. For example, using BCI technology to remotely control drones or driverless vehicles breaks distance limitations and allows operating various devices across spaces to complete different tasks, which will completely revolutionize the working modes and development of many industries.

At the same time, BCI technology can also be used for team collaboration, enabling seamless communication and cooperation among team members through brainwave signals, thereby improving work efficiency and collaboration effectiveness. This technology has broad application prospects in fields such as logistics, emergency rescue, and the military.

It can be said that when BCI technology meets wearable devices, a brand-new technological revolution is quietly unfolding. The combination not only endows transportation tools with unprecedented intelligent capabilities but also brings earth-shaking changes to life. We can foresee the future of life in the process of their integration and imagine an unparalleled new era.

## 13.4 Application of Brain–Computer Interface in Remote Driving

The goal of technological development must be to provide humans with more advanced and considerate services, and this logic is no exception in the transportation field. With the development of transportation technology, driving technology has gradually evolved from manual to automated, and automation is progressing toward unmanned operation. So what kind of transformation will occur beyond unmanned driving technology? According to the direction and logic of technological development, the upgrade of unmanned driving technology will move toward remote driving, because this direction can bring more convenience to our lives.

Imagine, when we get up in the morning to go to work, we only need to wear a head-mounted device and generate the idea of driving; our vehicle will automatically start and drive to a designated area to wait for us to get in. What a relaxed scene that would be. Or, when we get up in the morning to go to work, we only need to wear a head-mounted device, and public transportation vehicles will arrive punctually at the community entrance according to our expected departure time, waiting for us to board. What a pleasant scene that would be. Or, when we are busy in the office, we do not need to get up; we can operate the vehicle remotely to pick up and drop off children at school. What a comfortable scene that would be.

In fact, such a future is not far from us, because remote driving control technology has already achieved remarkable results. This technology has been practically applied in many fields such as industry and agriculture with the arrival of the 5G era. However, unmanned driving remains the mainstream of technological development in the transportation field, so the application of remote driving technology in transportation is still limited. But when BCI technology integrates with transportation technology, remote driving technology can demonstrate greater value. For example, the scenarios we imagined above can all be realized through the interaction of BCI and remote driving technologies.

By directly reading and decoding brain signals, BCI technology enables drivers to achieve higher accuracy and response speed when remotely controlling vehicles. The application of this technology is not limited to enhancing the driving experience but also shows great potential in work, life, entertainment, and other fields.

From a technical perspective, the future BCI technology can bring the following changes to remote driving technology, thereby enriching the application scenarios of remote driving technology.

**1. Improve the Accuracy and Response Speed of Remote Driving**

BCI technology can significantly enhance the accuracy and response speed of remote driving. High-sensitivity EEG sensors and advanced signal processing algorithms enable the system to capture and decode the driver's brainwave signals in real time, transmitting them to the remote driving system with almost no delay. This greatly reduces operational latency, ensuring that the driver's intentions are quickly and accurately converted into actual operations.

Moreover, with the advancement of wireless transmission technologies, especially the widespread application of 5G and future 6G networks, the transmission speed and stability of brainwave signals will be significantly improved. This will further enhance the real-time performance and reliability of remote driving, allowing drivers to achieve more precise and flexible control in complex driving environments.

**2. Operability and Flexibility of Remote Driving**

The progress of BCI technology will greatly enhance the operability and flexibility of remote driving. In the future, drivers will no longer need to rely on traditional physical control devices but will directly control remote vehicles through brainwave signals. This mode of operation not only improves the intuitiveness and accuracy of control but also enables drivers to respond more flexibly to various complex road conditions and emergencies.

As BCI technology develops, it will be able to adjust control strategies in real time based on the driver's brainwave signals, thereby enhancing the adaptive capability of remote driving systems. For example, in urban traffic, the BCI system can intelligently regulate speed, lane changes, and parking according to the driver's intentions; on highways, the system can automatically identify and avoid obstacles, improving driving safety and efficiency. This highly flexible mode of operation will significantly expand the application scope and practicality of remote driving.

**3. Virtual Training and Simulation in Remote Driving**

BCI technology will provide remote drivers with more advanced and comprehensive virtual training and simulation systems. Future driver training will not be limited to practical exercises in real environments but will combine virtual reality (VR) technology with BCI systems to offer drivers a highly immersive training experience.

In virtual training, drivers can directly participate in simulations of various complex road conditions and emergencies through the BCI system. This highly realistic training environment not only improves drivers' operational skills but also enhances their ability to respond in emergencies. Through continuous simulation practice, drivers can better master remote driving technology, improving safety and efficiency in actual operations.

**4. Group Control in Remote Driving**

It is believed that with the integration of BCI technology and remote driving technology, it will become possible for a single person to operate multiple vehicles simultaneously, and multiple vehicles can be controlled collaboratively. In some complex tasks, such as large-scale logistics transportation, emergency rescue, and military operations, operating a single vehicle may not meet the demands, while the group control function of BCI technology will provide an effective solution.

Additionally, through the BCI system, collaborative operation by multiple drivers can also be realized. For example, multiple drivers can jointly control one or more remote vehicles through brainwave signals, coordinating to complete complex operational tasks. This mode of group control and collaboration will significantly enhance the functionality and application scope of remote driving systems.

**5. Enhancing Social and Economic Impact**

With the widespread application of BCI technology in remote driving, profound changes will occur in social and economic aspects. First, BCI technology will significantly improve traffic safety, reduce the occurrence of traffic accidents, and lower social and economic costs. Second, BCI technology will promote the intelligent transformation of the transportation industry, spawning numerous emerging industries and employment opportunities, thereby boosting economic development.

At the social level, BCI technology will enhance people's travel experience, especially by providing more travel convenience for the elderly and people with disabilities. By controlling wheelchairs and assistive devices through brainwaves, people with disabilities can achieve more autonomous and free mobility, improving their quality of life and social participation.

In summary, in the future development of driving technology, remote driving will become a major focus. Driven by BCI technology, this driving mode will play an increasingly important role in the transportation field, leading the transportation industry toward a highly intelligent direction.

## 13.5 The Impact of Brain–Computer Interface Technology on the Transportation Industry and Society

Standing at the forefront of technology and looking toward the future of transportation development, it is hard not to be excited by the infinite possibilities brought by BCI technology. We can clearly foresee that in the near future, not only will we be able to control everything at home through smartphones, but we will also be able to drive our vehicles with the power of our minds. All of this is no longer a fictional scenario from science fiction movies but is becoming an attainable reality driven by BCI technology.

BCI technology achieves seamless interaction between humans and machines by reading and decoding brain signals, fundamentally changing our understanding and operation of vehicles, traffic management, and even the entire transportation system. This technology is enhancing the intelligence and safety of vehicles and will bring unprecedented economic benefits and social welfare. It will drive a comprehensive transformation and upgrade of the transportation industry, making our travel more convenient, efficient, and environmentally friendly, while bringing broader benefits to society.

In this era full of opportunities and challenges, BCI technology paints a grand blueprint for the future of transportation. Next, let us analyze the changes that BCI technology will bring to the transportation industry and society as a whole.

**1. Technological Advancement and Transformation of the Transportation Industry**

BCI technology is driving profound changes in the transportation industry. Traditional vehicles rely on physical controls and complex mechanical systems, whereas BCI technology enables direct control of vehicles through neural signals, fundamentally transforming the design and operation of transportation tools.

BCI technology enables more intelligent operation of vehicles. Through control by brainwave signals, drivers can operate vehicles with greater precision, reducing dependence on mechanical control devices. This intelligent mode of operation not only improves driving efficiency but also reduces the risk of human error. BCI technology also offers new solutions for autonomous driving. In autonomous driving systems, BCI technology can help the system quickly recognize the driver's intentions and respond accordingly, enhancing the safety and reliability of autonomous driving.

**2. Economic Benefits and Industry Development**

The widespread application of BCI technology has not only driven technological progress in the transportation industry but also brought significant economic benefits. By enhancing the intelligence level of vehicles, BCI technology has created new market opportunities and business models for the transportation sector.

Currently, the application of BCI technology has already spawned numerous innovative products and services. For example, brainwave-controlled devices such as smart helmets and smart seats are in growing demand, providing new growth points for both technology and economic development.

Moreover, the rapid development of BCI technology has also stimulated the rise of related industries. The research and production of core technologies such as brainwave sensors, signal processing chips, and intelligent control systems have created new business opportunities for related enterprises. At the same time, the application of BCI technology has promoted the development of training and service industries, such as increased demand for professional driving training and intelligent transportation consulting services, offering more room for industry growth.

**3. Social Welfare and Improvement of Quality of Life**

The application of BCI technology in the transportation industry has brought about not only technological and economic changes but also positive impacts on social welfare and quality of life.

BCI technology makes vehicle operation simpler and smarter, especially providing greater travel convenience for the elderly and people with disabilities.

In terms of environmental protection, this technology also promotes social sustainable development. By optimizing traffic flow and reducing traffic accidents, it lowers fuel consumption and carbon emissions from vehicles. BCI technology can enhance transportation development while reducing environmental pollution. Additionally, this technology can be combined with new energy technologies and intelligent driving technologies to promote green transportation and the realization of sustainable development.

BCI technology, as an innovative driving force in the transportation industry, is profoundly changing our travel methods and social life. By enhancing the intelligence level of transportation tools, BCI technology brings technological advancement, safety assurance, and economic benefits to the transportation industry, while also improving social welfare and quality of life. In the future, we look forward to seeing more innovations and breakthroughs that will lead the transportation industry toward a smarter, more efficient, and sustainable development path.

# Bibliography


A new generation of domestically produced dedicated acquisition chips for brain-computer interfaces was successfully developed in Tianjin! *Information Systems Engineering*, 2023, (11):2.

Almarzouki H Z, Alsulami H, Rizwan A, *et al.* An internet of medical things-based model for real-time monitoring and averting stroke sensors. *Journal of Healthcare Engineering*, 2021, 2021.

An Jialin, Zhang Shen, Wang Ling, *et al.* Comparative study of functional near-infrared spectroscopy and functional magnetic resonance imaging motor imagery brain–computer interface bimodal. *Biomedical Engineering Research*, 2018, 37(03):265–270.

An Tuo, Liang Hanyu, Zhou Huiling, *et al.* The relationship between cyberbullying, fear of negative evaluation, and social anxiety among college students in a university in Fujian Province. *Journal of Guizhou Medical University*, 2022, 47(09):1052–1056.

Andersen R A and Aflalo T. Preserved cortical somatotopic and motor representations in tetraplegic humans. *Current Opinion in Neurobiology*, 2022, 74: 102547.

Artificial Intelligence Medical Device Innovation Cooperation Platform. *White Paper on the Application of Brain-Computer Interface Technology in the Medical and Health field.* Beijing: Artificial Intelligence Medical Device Innovation Cooperation Platform, 2023.

Aurucci G V, Preatoni G, Damiani A, *et al.* Brain-computer interface to deliver individualized multisensory intervention for neuropathic pain. *Neurotherapeutics*, 2023: 1–14.

Bai Ruiyu, Wang Xuanqi, Zhang Zimo, *et al.* Current status and trends of cortical minimally invasive brain-computer interface sensors. *Information and Communication Technology and Policy*, 2024, 50(05):18–25.

Banach K, Małecki M, Rosół M, *et al.* Brain–computer interface for electric wheelchair based on alpha waves of EEG signal. *Bio-Algorithms and Med-Systems*, 2021, 17(3):165–172.

Bergeron D, Iorio-Morin C, Bonizzato M, *et al.* Use of invasive brain-computer interfaces in pediatric neurosurgery: Technical and ethical considerations. *Journal of Child Neurology*, 2023, 38(3-4):223–238.

Brain-computer interface technology. *Middle School Student Reading (Junior High Edition)*, 2024, (Z2):74–75.

Cao Chengxu, Qishisan, and Jintong Lin. The relationship between social self-efficacy and social anxiety among college students: The mediating role of fear of negative evaluation. *Journal of Neijiang Normal University*, 2021, 36(12):21–24.

Cervera M A, Soekadar S R, Ushiba J, *et al.* Brain-computer interfaces for post-stroke motor rehabilitation: a meta-analysis. *Annals of Clinical and Translational Neurology*, 2018, 5(5):651–663.

Cha Wei, Yang Yuxiao, and Hu Shengrong. Research on the development trends of brain-computer interface industry in Zhejiang Province. *Information Communication Technology and Policy*, 2024, 50(05):47–53.

Chen Jiaxi. *The Impact of Fear of Negative Evaluation on Employee Voice Behavior: The Roles of Social Anxiety and Organizational Identification.* Central University of Finance and Economics, 2022.

Chen Lebin. *Research on Frequency Modulated Visual Brain-Computer Interface Based on Transfer Learning*. Nanchang University, 2021.

Chen Tao. Reflections and direction suggestions on domestic brain-computer interface development. *Shanghai Informatization*, 2024, (06):14–17.

Cheng Liwei, Zhang Qian, Liang Liyan, *et al.* High-speed communication demands and related application scenarios for brain-computer interface. *Information and Communication Technology and Policy*, 2024, 50(05):12–17.

Colucci A, Vermehren M, Cavallo A, *et al.* Brain–computer interface-controlled exoskeletons in clinical neurorehabilitation: Ready or not? *Neurorehabilitation and Neural Repair*, 2022, 36(12):747–756.

Ding Lingsong. *Research on UAV Control System Based on Portable Electrooculography Acquisition Device and Hybrid Brain-Computer Interface*. South China University of Technology, 2021.

Dong Yue, Liu Ke, and Wang Tao. Application of hybrid-modal brain-computer interface technology: A new direction in neural rehabilitation. *Information and Communication Technology and Policy*, 2024, 50(05):41–46.

Du Shuo. *Research on Motor Imagery Brain-Computer Interface for Intelligent Assisted Driving*. Yanshan University, 2019.

Du Wentao. *Research on Automotive Cabin Sound and Vibration Experience Based on Brain-Computer Interface*. Southwest Jiaotong University, 2022.

Fang Cheng. *Research on Phase Shift Value Optimization Strategy Based on Brain-Computer Interface*. Xidian University, 2020.

Flesher S N, Downey J E, Weiss J M, *et al.* A brain-computer interface that evokes tactile sensations improves robotic arm control. *Science*, 2021, 372(6544):831–836.

Fu Yunfa, Gong Anmin, and Nan Wenya. *Principles and Practice of Neurofeedback*. Beijing: Electronics Industry Press, 2021: 33–34.

Fu Yunfa, Gong Anmin, Chen Chao, *et al. Practical Brain-Computer Interface: Narrowing the Gap Between Research and Real-World Applications*. Beijing: Publishing House of Electronics Industry, 2022: 45–47.

Fu Yunfa, Wang Fan, Ding Peng, *et al. Brain-Computer Interface*. Beijing: National Defense Industry Press, 2023: 631–646.

Fu Yunfa, Yang Qiuhong, Xu Baolei, *et al. Principles and Practice of Brain-Computer Interface*. Beijing: National Defense Industry Press, 2017: 5–6.

Fuyun Fu, Yanlong Guo, Xiabing Zhang, *et al. Brain-Computer Interface—A Revolutionary Human-Computer Interaction*. Beijing: National Defense Industry Press, 2020: 52–56.

Gan Lihao. How brain-computer interface technology reshapes human communication. *Social Science News*, 2024-06-27: 006.

Gao Xiaorong and Li Wenyu. Special topic introduction: Brain-computer interface. *Information and Communication Technology and Policy*, 2024, 50(05):1.

Guo Yi. *Remote Control and Implementation of UAV Based on ROS and EEG*. Xidian University, 2018.

He Jing. Brain-computer interface does not constitute a challenge to embodied cognition. *Journal of Shanghai Normal University (Philosophy and Social Sciences Edition)*, 2024, 53(03):41–48.

He Liushi, Xie Jun, Yu Hongwei, *et al.* Design of a steady-state visual evoked potential brain-computer interface system integrating eye tracking and dynamically adjustable targets. *Journal of Xi'an Jiaotong University*, 2021, 55(10):87–95.

He Qinghua. *Experimental Research on Brain-Computer Interface Based on Visual Evoked Potentials*. Chongqing University, 2003.

Hong Hengfei and Jiang Yun. Brain-computer interface technology enables paraplegic patients to "think" words. *Science and Technology Daily*, 2024-05-23: 006.

Hramov A E, Maksimenko V A, and Pisarchik A N. Physical principles of brain–computer interfaces and their applications for rehabilitation, robotics and control of human brain states. *Physics Reports*, 2021, 918:1–133.

Hu Ling. Understanding the general model of technological regulation: Taking brain-computer interface as an example. *Eastern Jurisprudence*, 2021, (04):38–48.

Huang Mingfen. From new media, sci-fi creativity to general artificial intelligence: The artistic revolution of brain-computer interface. *Journal of Ocean University of China (Social Sciences Edition)*, 2024, (04):109–118.

Huang Xiaolong, Chen Xinyu, and Li Xuefeng. Legal regulation prospects applicable to brain-computer interface technology in investigation. *Journal of Zhejiang Police College*, 2024, (03):113–124.

Huang Ying, Bian Xinyi, Gao Peng, *et al.* Effects of brain-computer interface technology on balance function and serum IL-6, TNF-α levels in patients with ischemic stroke. *Journal of Anhui Medical University*, 2024, 59(07): 1263–1268.

Huang Zhe. Brain-computer interface changes life. *China Computer Newspaper*, 2024-05-13: 010.

Jiang Jun. *Research on Adaptive Brain-Computer Interface Control System.* National University of Defense Technology, 2011.

Jiang Yanli and Liu Xinyu. Empirical investigation of brain-computer interface in classroom teaching. *Internet Weekly*, 2024, (10):50–53.

Jovanovic L I, Kapadia N, Zivanovic V, *et al.* Brain–computer interface-triggered functional electrical stimulation therapy for rehabilitation of reaching and grasping after spinal cord injury: A feasibility study. *Spinal Cord Series and Cases*, 2021, 7(1): 24.

Klein E and Ojemann J. Informed consent in implantable BCI research: Identification of research risks and recommendations for development of best practices. *Journal of Neural Engineering*, 2016, 13(4): 043001.

Klein E, Brown T, Sample M, *et al.* Engineering the brain: Ethical issues and the introduction of neural devices. *Hastings Center Report*, 2015, 45(6):26–35.

Klein E. Ethics and the emergence of brain-computer interface medicine. *Handbook of Clinical Neurology*, 2020, 168:329–339.

Lan Zhen, Li Zixing, Yan Chao, *et al.* Research review on UAV control system based on brain-computer interface. *Control Theory and Applications*, 2023, 40(12):2142–2159.

Li Jiawei, Zhang Hongxin, and Xu Ruilin. Design of UAV formation control system based on brain-computer interface. *Aeronautical Science and Technology*, 2023, 34(02):104–110.

Li Jiawei. *Simulation and Implementation of UAV Formation Control System Based on Brain-Computer Interface.* Beijing University of Posts and Telecommunications, 2023.

Li Jiefang. *Research on Embedded RSVP System Based on Brain-Computer Interface.* Beijing Jiaotong University, 2022.

Li Xiaoyong. Logical justification and institutional conception of mental integrity rights under brain-computer interface technology background. *Political and Legal Forum*, 2024, (03):45–57.

Li Yan. *Research on Motor Imagery Brain-Computer Interface Based on Transfer Learning.* Nanchang University, 2021.

Li Yu. *Research on fNIRS-Based Brain-Computer Interface System.* Kunming University of Science and Technology, 2020.

Liang Jingkun. *EEG Signal Analysis and Brain-Computer Interface Research Based on Imagined Driving Behavior.* Hebei University of Technology, 2012.

Liang Liyan, Kong Shuyi, Zhang Qian, *et al.* Discussion and research on paradigms, algorithms, and encoding-decoding concepts of brain-computer interface. *Information Communication Technology and Policy*, 2024, 50(05): 61–70.

Liu Chang. *Research on Steady-State Visual Evoked Potential Recognition Algorithm for Practical Brain-Computer Interface.* Xiamen University, 2021.

Liu Guangming. A philosophical examination of brain-computer interface replacing the body's "limits". *Medicine and Philosophy*, 2024, 45(09):11–15.

Liu Kunjia. *Optimization Algorithm of Visual Evoked Potential Brain-Computer Interface and Its Application.* National University of Defense Technology, 2018.

Liu Kunjie. *Research on the Influence of Fear of Negative Evaluation on Social Anxiety of Junior High School Students and Its Intervention.* Huaibei Normal University, 2022.

Liu Shengjie. *Research on Brain-Computer Interface Decoding Algorithms Based on Stereoscopic EEG.* Shanghai Jiao Tong University, 2021.

Liu Shimeng. Accelerated construction of brain-computer interface standards in China. *Huaxia Times*, 2024-07-15: 006.

Liu Siyu, Zhang Deyu, Ming Zhiyuan, *et al.* Remote brain-controlled UAV system based on brain-computer interface and human-machine closed loop. *Acta Armamentarii*, 45(9), 3191–3203.

Liu Weiqing. *Research on a Combined Acquisition System Based on EEG-NIRS Brain-Computer Interface.* Nanjing University of Aeronautics and Astronautics, 2021.

Liu Yadong, Li Ming, and Zhou Zongtan. Military application prospects and challenges of brain–computer interface technology. *Control Theory & Applications*, 41(11), 1991–2001.

Liu Yaru. *Research on Multi-Mobile Target Selection Technology Based on Brain-Computer Interface.* National University of Defense Technology, 2016.

Liu Zhaojun. *Transfer Learning Brain-Computer Interface for Intelligent Assisted Driving.* Yanshan University, 2020.

Liu Zhihong, Zhang Ye, Zhao Zhiming, *et al.* The relationship between fear of negative evaluation and excessive internet use among college students: The sequential mediating role of social anxiety and self-compensation motivation. In *Proceedings of the 24th National Psychology Academic Conference.*

Chinese Psychological Society. School of Educational Science, Shenyang Normal University, 2022:2.

Long Sheng. *Vehicle Static Function Control Technology Based on Brain-Computer Interface*. National University of Defense Technology, 2019.

Lu Xiaobin and Li Zhijun. Brain-computer interface technology leading a new path for intelligent warfare. In *Proceedings of the 12th China Command and Control Conference (Volume 1)*. Chinese Institute of Command and Control. Information Engineering University; PLA Unit 61705, 2024:5.

Lü You. Criminal risk analysis of brain-computer interface technology applications under the background of artificial intelligence. In *Collection of Shanghai Legal Studies (Vol. 5, 2020, Total Vol. 29) — Proceedings of the 2020 World Artificial Intelligence Conference Rule of Law Forum*. World Artificial Intelligence Conference Organizing Committee, Shanghai Law Society. Qingpu District People's Procuratorate of Shanghai, 2020:8.

Lyu X, Ding P, Li S, *et al.* Human factors engineering of BCI: An evaluation for satisfaction of BCI based on motor imagery. *Cognitive Neurodynamics*, 2023, 17(1):105–118.

Lyu Xiaotong, Ding Peng, Li Siyu, *et al.* Human factors engineering and applications of brain-computer interfaces: User-centered design and evaluation methods for BCIs. *Journal of Biomedical Engineering*, 2021, 38(2):210–223.

Ma Cong, Zhang Yufang, Qin Rongzhen, *et al.* Research progress on fear of negative evaluation among college students. *Occupational and Health*, 2024, 40(07):1000–1004.

Ma Cong, Zhang Yufang, Qin Rongzhen, *et al.* The relationship between college students' social anxiety and mobile phone addiction: The parallel mediation effect of technological intrusion and fear of negative evaluation. *Chinese Journal of Health Psychology*, 2024, 32(06):926–933.

Ma Jun. How far is brain-computer interface from "human-machine symbiosis". *Global Times*, 2024-06-04: 008.

Ma Y, Gong A, Nan W, *et al.* Personalized brain–computer interface and its applications. *Journal of Personalized Medicine*, 2022, 13(1): 46.

Matarasso A K, Rieke J D, White K, *et al.* Combined real-time fMRI and real-time fNIRS brain computer interface (BCI): Training of volitional wrist extension after stroke, a case series pilot study. *PLoS One*, 2021, 16(5): e0250431.

McGeady C, Vučković A, Singh Tharu N, *et al.* Brain-computer interface priming for cervical transcutaneous spinal cord stimulation therapy: An exploratory case study. *Frontiers in Rehabilitation Sciences*, 2022, 3: 896766.

Meng Fei. *Recognition Methods and Applications of Motor Imagery EEG Signals in Brain-Computer Interface*. Dalian Jiaotong University, 2013.

Meng Lixia, Tao Linmi, Sun Fuchun, *et al.* Mobile vehicle navigation system based on brain-computer interface and dual LiDAR. *Robot*, 2012, 34(04): 449–454+459.

Merk T, Peterson V, Lipski W J, *et al.* Electrocorticography is superior to subthalamic local field potentials for movement decoding in Parkinson's disease. *Elife*, 2022, 11: e75126.

Metzger S L, Littlejohn K T, Silva A B, *et al.* A high-performance neuroprosthesis for speech decoding and avatar control. *Nature*, 2023, 620(7976): 1037–1046.

Musk's brain-computer interface experiment shows progress again (192). *Journal of Clinical Electrocardiology*, 2024, 33(03):234.

Nenadic Z. Brain–computer interfaces for human gait restoration. *Control Theory and Technology*, 2021: 1–13.

Pais-Vieira C, Gaspar P, Matos D, *et al.* Embodiment comfort levels during motor imagery training combined with immersive virtual reality in a spinal cord injury patient. *Frontiers in Human Neuroscience*, 2022, 16: 909112.

Pan Xin. *Virtual Driving System Based on Brain-Computer Interface*. South China University of Technology, 2017.

Qian Tongxin. Accelerated development of brain-computer interface technology in China moving from laboratory to market. *First Financial Daily*, 2024-07-03: A01.

Qin Changsen and Liu Shuting. Empowerment, real risks, and compliance protection of brain-computer interfaces in intelligence activities. *Science and Technology Review*, 2025, 43(4):67–74.

Qin Weili. Brain-computer interface is accelerating "access" to health. *Shanxi Elderly*, 2024, (06):10–12.

Song Jing. The amazing brain-computer interface. *Everyone's Health*, 2024, (16):18.

Sun Tingyang and Jia Xuan. Brain-computer interface, quantum computing, commercial space... Understanding future industries and seeing a wonderful future. *China Economic Weekly*, 2024, (09):50–56.

Tang Jingsheng. *Research on Coordinated Control Technology Based on Brain-Computer Interface*. National University of Defense Technology, 2013.

Tuo Yan and Zhang Wenxin. "Brain-computer interfaces" ready to take off. *Guangdong Science and Technology News*, 2024-07-05: 002.

Vansteensel M J, Branco M P, Leinders S, *et al.* Methodological recommendations for studies on the daily life implementation of implantable communication-brain–computer interfaces for individuals with locked-in syndrome. *Neurorehabilitation and Neural Repair*, 2022, 36(10-11):666–677.

Vansteensel M J, Klein E, van Thiel G, *et al.* Towards clinical application of implantable brain–computer interfaces for people with late-stage ALS: Medical and ethical considerations. *Journal of Neurology*, 2023, 270(3): 1323–1336.

Wang Bing, Li Sha, Wang Wenjie, *et al.* The influence of college students' self-identity on social avoidance and distress: The mediating effect of fear of negative evaluation. *Journal of China Medical University*, 2021, 50(12): 1107–1111.

Wang Jingjun. Research on somatosensory P300 brain-computer interface paradigm. National University of Defense Technology, 2016.

Wang Ning, Sun Guangjin, and Liu Xuewen. Design of a helmet display/aiming system based on brain-computer interface. *Application of Electronic Technology*, 2015, 41(05):149–151+155.

Wang Sanqiang. Design of EEG signal acquisition system and its application in brain-computer interface. Chongqing University, 2006.

Wang Shuo and Shi Sishi. What is brain-computer interface. *Invention and Innovation (Elementary School Students)*, 2024, (05):35.

Wang Weihua. *Research on Neural Signal Decoding of implantable Brain-Computer Interface Based on Machine Learning*. Huazhong University of Science and Technology, 2021.

Wang Yuding. *Design and Research of Real-Time Brain-Computer Interface*. Chongqing University, 2012.

Wang Z, Shi N, Zhang Y, *et al.* Conformal in-ear bioelectronics for visual and auditory brain-computer interfaces. *Nature Communications*, 2023, 14(1): 4213.

Wei Mengjia and Song Chen. Brain-computer interface, how many possibilities does it connect to the future? *Xinhua Daily Telegraph*, 2024-04-30: 006.

Wei Sisong. *Research on Aircraft Virtual Reality Simulation Driving System Based on Brain-Computer Interface*. Nanjing University of Aeronautics and Astronautics, 2021.

Willett F R, Avansino D T, Hochberg L R, *et al.* High-performance brain-to-text communication via handwriting. *Nature*, 2021, 593(7858):249–254.

Willett F R, Kunz E M, Fan C, *et al.* A high-performance speech neuroprosthesis. *Nature*, 2023, 620(7976):1031–1036.

Wu Chenyang. *The Impact of Fear of Negative Evaluation on Social Anxiety: The Moderating Role of Cognitive Reappraisal and Intervention Study*. Jiangxi Normal University, 2023.

Wu Qiaojun. Laying out new tracks for brain-computer interfaces to strive for international status. *Tianjin Daily*, 2024-06-21: 007.

Xia Miaoyun, Wang Ying, Yang Qianzi, *et al.* Application of brain-computer interface in neuroscience. *Shanghai Medical Journal*, 2023, 46(05):278–281.

Xiamen University of Technology cooperated with well-known enterprises to hold an academic forum on wearable brain-computer interface devices. *Journal of Xiamen University of Technology*, 2024, 32(03):80.

Xu Gengyou. *Research on Classification and Recognition of Motor Imagery EEG Signals in Brain-Computer Interfaces*. Xi'an University of Technology, 2024.

Xu Jiang. Wireless remote control car system based on real-time brain-computer interface. Chongqing University, 2010.

Xu Ming. Ethical challenges of brain-computer interface technology and its rational regulation research. *Information and Communication Technology and Policy*, 2024, 50(05):54–59.

Xu Siyu. *The Relationship and Intervention Study of Fear of Negative Evaluation, Coping Styles, and Social Anxiety Among Junior High School Students.* Central China Normal University, 2022.

Xue H, Wang D, Jin M, *et al.* Hydrogel electrodes with conductive and substrate-adhesive layers for noninvasive long-term EEG acquisition. *Microsystems & Nanoengineering*, 2023, 9(1): 79.

Yang Banghua. *An Emotiv Brain-Controlled UAV System and Method based on VC++ and Matlab Hybrid Programming.* Shanghai University, 2018-12-01.

Yang Guang. Brain-computer interface industry expected to receive more "policy catalysis". *China Informatization Weekly*, 2024-07-08: 015.

Yang Qingfeng. Brain-computer interface, forms of life, and identity. *Journal of Donghua University (Social Sciences Edition)*, 2024, 24(02):26–31. DOI:10.19883/j.1009-9034.2024.0096.

Yang Shu. *The Impact of Fear of Negative Evaluation on College Students' Mobile Phone Dependence: The Mediating Role of Loneliness and Positive Self-Presentation*. Qufu Normal University, 2021.

Yang Xin, Wu Bian, Chen Weidong, *et al.* Potential applications of brain-computer interface technology in the aerospace field. *Manned Spaceflight*, 2012, 18(03):87–92.

Yang Yanfeng. *The Impact of Life Events on Social Anxiety of Left-Behind High School Students: The Mediating Role of Fear of Negative Evaluation and the Moderating Role of Friend Support*. Xinyang Normal University, 2022.

Yu Guifang, Wang Haiyun, Li Xiaojian, *et al.* Comparative study on the technological competition pattern between China and the US in the field of brain-computer interface based on bibliometrics and patent analysis. *Science and Technology Management Research*, 2024, 44(03):153–162.

Yu Qinwen, Zhou Wangcheng, Dai Yakang, *et al.* A review of motor imagery brain-computer interface encoding and decoding algorithms in embedded systems. *Computer Engineering and Applications*, 2024, 60(18):50–65.

Yu Shuyue, Li Xiang, Yu Gongjing, *et al.* Development and prospects of brain-computer interface technology. *Computer Measurement and Control*, 2019, 27(10):5–12.

Yu Yang. *Research on Asynchronous Brain-Computer Interface Control Technology for Mobile Platforms*. National University of Defense Technology, 2017.

Yuan Jiping, He Yuan, and Chang Qing. Preliminary exploration of interdisciplinary brain-computer interface curriculum in junior high school—Taking the brain-computer interface principles course as an example. *Teaching Management and Educational Research*, 2024, 9(09):14–18.

Zhang Lulin. *Research on Control and Virtual Scene Application Based on Brain-Computer Interface*. Harbin Engineering University, 2013.

Zhang Mengran. New technology enhances non-invasive brain-computer interface functions. *Science and Technology Daily*, 2024-06-20: 004.

Zhang Miao. How far are "brain-computer interfaces" from daily life with frequent updates. *China Youth Daily*, 2024-07-05: 003.

Zhang Qingqing, Chen Tingting, Liu Shanshan, *et al.* The impact of fear of negative evaluation on depression: Multiple sequential mediation effects of cognitive reappraisal, expressive suppression, and social anxiety. *Chinese Journal of Clinical Psychology*, 2022, 30(06):1292–1296.

Zhang Ruilong. *Research on Multi-UAV Brain-Like Intelligent Task Decision-Making and Motion Planning Methods in Complex Environments*. Tianjin University, 2022.

Zhang Siwei and Zhao Jizong. Brain-computer interface human trials must be cautiously promoted. *Medical Science News*, 2024-06-28: 002.

Zhang Ying and Zhang Bowen. Current status and development of brain-computer interface technology in medical rehabilitation — Interview with Professor Fan Yubo from Beihang University. *Micro-Nano Electronics and Intelligent Manufacturing*, 2022, 4(03):2–4.

Zhang Yiwen. AI storm sweeps the automotive industry. *China Automotive News*, 2024-03-04: 040.

Zhang Yue. *The Relationship Between Peer Rejection and Social Anxiety in Junior High School Students: The Role of Fear of Negative Evaluation and Self-Compassion and Educational Implications*. Henan University, 2022.

Zhang Zhe, Zhao Xu, Ma Yixin, *et al.* Ethical considerations of brain-computer interface technology. *Journal of Biomedical Engineering*, 2023, 40(2): 358–364.

Zhang Zhiyao and Wang Xueyan. A study on the differences in social appearance anxiety between psychology and non-psychology major college students: The mediating role of fear of negative evaluation. *Campus Psychology*, 2023, 21(06):449–452.

Zhao Xiujuan. *Research and Application of UAV Control Based on Brain-Computer Interface*. Qufu Normal University, 2020.

Zheng Juntong, Chen Huixin, Yang Shuang, *et al.* The impact of fear of positive evaluation on social anxiety: The mediating role of fear of negative evaluation. *Psychological Monthly*, 2023, 18(23):38–40.

Zhi Ang. The early stage of brain-computer interface technology explosion: What is the value of "Bei Nao No. 2"? *Computer Newspaper*, 2024-05-13: 020.

Zhong Yingzhu. *Spatial Cognition Training System Integrating Brain-Computer Interface and Virtual UAV and its EEG Analysis*. Yanshan University, 2022.

Zhong Zhe and Wu Yanan. Bringing "brain-computer interface" into daily life. *Southern Daily*, 2024-05-24: A07.

Zhou Chun, Yang Chaoxu, and Rong Haijun. Multimodal brain-computer interface technology for quadcopter flight control. *Flight Mechanics*, 2021, 39(04):81–87.

Zhou Junyu, Liu Yaqun, Zhang Xueliang, *et al.* Design and implementation of a multi-feature fusion UAV control system. *Electronic Design Engineering*, 2023, 31(16):1–5.

Zhu M, Chen J, Li H, *et al.* Vehicle driver drowsiness detection method using wearable EEG based on convolutional neural network. *Neural Computing and Applications*, 2021, 33(20):13965–13980.

Zhu Pengjin, Huang Nanxiang, and Wang Hao. Application prospects and challenges of brain-computer interface technology in military intelligence. In *Proceedings of the 12th China Command and Control Conference (Volume 1)*. Chinese Institute of Command and Control. School of International Relations, National University of Defense Technology, 2024:5.

# Index

**E**